Logic, Argumentation & Reasoning

Interdisciplinary Perspectives from the Humanities and Social Sciences

Volume 27

Logic, Argumentation & Reasoning (LAR) explores links between the Humanities and Social Sciences, with theories (including decision and action theory) drawn from the cognitive sciences, economics, sociology, law, logic, and the philosophy of science.

Its main ambitions are to develop a theoretical framework that will encourage and enable interaction between disciplines, and to integrate the Humanities and Social Sciences around their main contributions to public life, using informed debate, lucid decision-making, and action based on reflection.

- Argumentation models and studies
- Communication, language and techniques of argumentation
- Reception of arguments, persuasion and the impact of power
- Diachronic transformations of argumentative practices

LAR is developed in partnership with the Maison Européenne des Sciences de l'Homme et de la Société (MESHS) at Nord - Pas de Calais and the UMR-STL: 8163 (CNRS).

This book series is indexed in SCOPUS.

Proposals should include :

- A short synopsis of the work, or the introduction chapter
- The proposed Table of Contents
- The CV of the lead author(s)
- If available: one sample chapter

We aim to make a first decision within 1 month of submission. In case of a positive first decision, the work will be provisionally contracted—the final decision about publication will depend upon the result of an anonymous peer review of the complete manuscript.

The complete work is usually peer-reviewed within 3 months of submission.

LAR discourages the submission of manuscripts containing reprints of previously published material, and/or manuscripts that are less than 150 pages / 85,000 words.

For inquiries and proposal submissions, authors may contact the editor-in-chief, Shahid Rahman at: shahid.rahman@univ-lille.fr, or the managing editor, Juan Redmond, at: juan.redmond@uv.cl

Tanja Rechnitzer

Applying Reflective Equilibrium

Towards the Justification of a Precautionary Principle

Springer

Tanja Rechnitzer
Institute of Philosophy
Leibniz University Hannover
Hannover, Germany

This work was supported by Swiss National Science Foundation.

ISSN 2214-9120 ISSN 2214-9139 (electronic)
Logic, Argumentation & Reasoning
ISBN 978-3-031-04332-1 ISBN 978-3-031-04333-8 (eBook)
https://doi.org/10.1007/978-3-031-04333-8

This Springer imprint is published by the registered company Springer Nature Switzerland AG
The registered company address is: Gewerbestrasse 11, 6330 Cham, Switzerland

Acknowledgements

This book is the revised version of my dissertation, which was defended at the University of Bern in 2018. Both while writing the dissertation and during the process of rewriting it, I profited immensely from the support and feedback of many people. Throughout the whole time, Claus Beisbart and Georg Brun supported me with invaluable advice, insightful feedback, and constructive criticism. The final manuscript benefited a great deal from continued discussions with Georg. A heartfelt thank you to both—I could not have wished for a better team of supervisors.

During a research stay at Harvard University, I was able to have repeated discussions with Catherine Z. Elgin, and her feedback greatly improved my work. I am extremely grateful for the constructive criticism and inspiring discussions. Her comments on the revised manuscript helped to make it both clearer and stronger.

A very special thank you goes to David Hopf for discussions, feedback, and support throughout the whole time I was working on this project. It was tremendously helpful, and it means a lot.

I am also grateful to Michael Schmidt for his detailed and helpful comments, and for many an insightful exchange.

I also want to thank the following people: Baptiste Le Bihan, Rodrigo Díaz, Andreas Freivogel, Kevin Helms, and Matthias Rolffs gave valuable feedback on parts of the revised manuscript. Dominik Aeschbacher, Annabel Colas, Philipp Emch, Daniel Liebeherr, Johanna Privitera, and Dominic Roser read individual chapters of the dissertation, offering constructive criticism and enabling me to resolve many problems, some small and some big. I also benefited from several discussions with Gregor Betz. Thanks also to the two anonymous reviewers of the manuscript for their constructive feedback.

I am grateful to have been able to conduct my research in a supportive and enriching environment. Many thanks to the people at the Institute of Philosophy at the University of Bern, specifically the PhD group and discussants at various *jour fixe* and round table meetings. A big thank you also to my new colleagues at Leibniz University Hannover, whose feedback helped me to finalize this book. An embarrassed but no less sincere thank you also goes to those I have inadvertently forgotten to list.

Both the Rechnitzer and Blatter families have been a great source of strength and support, as have my friends, and I am grateful to have so many wonderful people in my life. Sandra, thank you for all the encouragement, and for your unwavering belief in me. Elli, this may not be the kind of book you had in mind when you predicted I would mention you in my acknowledgments one day, but here you go.

And, of course, my deepest and fondest thanks go to Jonas Blatter. Jonas, I am glad I got to do this with you by my side. Thank you for the ongoing and inspiring discussions, all the feedback and tech-support—but thank you even more for everything else.

Last but not least, my thanks go to the Swiss National Science Foundation for making this project possible by their generous financial support for the open access publication of this book, and for funding my research as part of the project "Reflective Equilibrium—Reconception and Application" (grant no. 150251).

Contents

Chapter 1
Introduction

How should we approach uncertain threats of potentially very serious harm? For example, how long should social distancing measures be enforced against the spread of COVID-19? Should we research and develop climate engineering technologies as a measure against climate change harms? Should glyphosate herbicides be banned? Such and similar decisions have potentially far-reaching consequences for the environment or human health; yet they often have to be made under considerable uncertainty, for example, uncertainty about the extent of possible harm, its likelihood, or cause-and-effect relations. Frequently, precautionary principles (PPs) are proposed as an answer to such challenges, telling us that we have to act to prevent harm even if it is uncertain. However, this idea also comes in for criticism as being alarmist, anti-scientific, and in effect doing more harm than good by causing high costs and stifling innovation. The question of how we should deal with uncertain harms is clearly a controversial one. When we seek to address this issue, we are not only faced with the question of *whether* precautionary principles are justified. More fundamentally, the methodological question arises of *how* such principles can be justified—what is an adequate method for the justification of a precautionary principle?

One method that is often recommended for justifying principles is reflective equilibrium (RE). Appeals to RE are made in a wide range of philosophical disciplines, from bioethics to political philosophy to logic. Its basic idea—that systematic principles are justified through a process of mutual adjustments to our existing judgments about relevant cases—is readily recited, and it is often seen as a, or even the, method of philosophy (Lewis 1983, Introduction; Scanlon 2003, 149). However, implementations of RE as they can be found in the literature typically either use a sketchy conception of RE, are restricted to simplified cases, focus only on some particular elements of RE, or do not make the application explicit and traceable. This makes it difficult to critically evaluate the method, to assess its potential, and to engage with the criticism directed at it. Moreover, it leaves

T. Rechnitzer, *Applying Reflective Equilibrium*, Logic, Argumentation & Reasoning 27, https://doi.org/10.1007/978-3-031-04333-8_1

researchers who want to use the method at a loss: **how does one apply reflective equilibrium?**

This book provides an explicit and detailed case study for an application of the method of RE. It adopts an elaborate conception of the method and tests whether it can be used to develop and defend a precautionary principle. With respect to RE, I show that there is at least one sufficiently fleshed-out RE conception that can successfully be applied to actual and complex cases. With respect to precautionary principles, the case study demonstrates how a rights-based precautionary principle can be constructed and defended. By focusing in particular on RE as a method for the justification of a precautionary principle, the case study also addresses questions of methodology that so far have been neglected in the debate about precautionary principles. In this way, the book can simultaneously illuminate two different debates.

In this first chapter, I identify desiderata for a case study by giving an overview of the main ideas of reflective equilibrium and surveying existing applications of the method in Sect. 1.1. Section 1.2 introduces the debate about precautionary principles, and explains why this topic was chosen for the case study. Lastly, Sect. 1.3 gives an overview of this book, outlining the chapters in order to show how the identified tasks are addressed.

1.1 Reflective Equilibrium: Main Ideas and Previous Applications

I start by describing the main ideas of reflective equilibrium in Sect. 1.1.1. In Sect. 1.1.2, I give an overview of existing applications of reflective equilibrium, before identifying desiderata for a case study in Sect. 1.1.3.

1.1.1 Main Ideas of Reflective Equilibrium

At the core of reflective equilibrium we have the idea that two sets of elements have to be mutually adjusted with respect to each other: on the one hand, our commitments about a subject matter, and on the other hand, a systematic account of them, for example in the form of principles or a theory. RE thus takes our existing judgments and beliefs as the starting points of theorizing, but without treating them as fundamental fixed-points. It thereby takes into account that our judgments and beliefs are usually not accepted simply at random, but have at least some minimal credibility via the fact that we are committed to them and act on them in our daily practices.

By requiring that our commitments have to be brought into coherence with a systematic account of them, RE also acknowledges that our judgments might be wrong, or based on biases and prejudices. Trying to fit them into a coherent position with systematic elements can reveal that some of our judgments are not as credible as they seemed on their own. Conversely, it might turn out that statements which are not very credible in isolation can be combined to form a convincing picture.[1]

There are at least two major traditions of RE in philosophy: one builds primarily on the works of Rawls, which is why it is often called a “Rawlsian method”. Rawls (1971) coined the term “reflective equilibrium”, and under this name the method gained importance especially in moral and political philosophy. The other tradition goes back to Goodman (1983), and, originating in questions of theory choice and justification in philosophy of science and logic, sees RE less as a method for a specific task and rather develops a general RE-based epistemology. Nonetheless, there is a lot of overlap in how RE is conceptualized in these two traditions—arguably, Rawlsian conceptions of RE can be seen as particular specifications of the broader RE epistemology. In both traditions, RE is typically understood as an account of justification, which does not necessarily mean that it is truth-conducive (cf. Rawls 1974).[2]

In the Rawlsian tradition (Daniels 1979; Rawls 1971, 2001; Scanlon 2003), RE is often seen as being specifically concerned with the justification of principles of justice that can be accepted by all citizens in a pluralistic society. This made RE a prominent and influential method in political philosophy (e.g., De-Shalit and Wolff 2007). But the potential of the method for moral questions more broadly was quickly recognized, and variants of it can be found in moral epistemology (DePaul 1993; Tersman 1993) as well as in applied ethics (Beauchamp and Childress 2013; Doorn 2010b; Van der Burg and Van Willigenburg 1998b; Van Thiel and Van Delden 2010). Especially in bioethics, a conception of RE that includes empirical data has gained influence as an approach to problem solving in specific cases (de Vries and van Leeuwen 2010).

In the epistemology tradition, the idea to use reflective equilibrium as the basis of a more comprehensive epistemology (Goodman 1951, 1978, 1983; Scheffler 1954) has been most thoroughly explored by Elgin (1996, 2014, 2017). By building on the works of Goodman and Elgin, and connecting them with methodological ideas from Carnap like the method of explication, Brun (2013, 2016, 2020) and Baumberger and Brun (2017, 2021) have developed one of the most elaborate conceptions of reflective equilibrium to date. I describe this RE conception in more detail in Chap. 2, as I adopt it as the basis for testing the applicability of RE in a case study. What is particular about this conception is that it distinguishes between two components of the equilibrium process. On the one hand, we have to bring commitments concerning the subject matter into agreement with a systematic

[1] Compare the example of the theft of a textbook described by Elgin (2017, 69): we have reasons to be wary of the credibility of each witness, yet their testimonies—properly weighted—combine to a plausible account of what happened.

[2] Thus, RE neither presupposes nor precludes (moral, scientific, ...) realism. For a discussion of the connection between RE, epistemic justification, and truth, see Tersman (1993, 94–114).

account of these commitments—e.g., principles, laws, or a theory. On the other hand, this agreement between commitments and a system has to be balanced against further demands.

The first sense of "reflective equilibrium" concerns the agreement between commitments and a systematic account. This is achieved through a process of mutually adjusting both sides, with neither commitments nor the systematic account being privileged or safe from revision. This is the idea of "balancing" that is usually associated with reflective equilibrium, and fits with how Goodman characterizes the process of justification in the famous quote from *Fact, Fiction, and Forecast*:

> [Rules] and particular inferences alike are justified by being brought into agreement with each other. *A rule is amended if it yields an inference we are unwilling to accept; an inference is rejected if it violates a rule we are unwilling to amend.* The process of justification is the delicate one of making mutual adjustments between rules and accepted inferences, and in the agreement achieved lies the only justification needed for either. (Goodman 1983, 64, italics in original)

Concerning the second reading of the metaphor "reflective equilibrium", (Brun 2020) argues that the agreement between commitments and a system has to be sought between two further forces that drive the process of adjustments in RE. Commitments should not only be in agreement with a system, but also have to respect the initial commitments that we started out with; and the system should not only be able to account for the commitments, but also has to do justice to certain theoretical desiderata which drive the theoretical development in the context at hand (Brun 2020, 26). In other words, the commitments that result after the process of adjustments should still be reasonable in light of the commitments that we started out with. This means that if an initial commitment was revised, we need to be able to give a plausible explanation for this revision. This is also central to Elgin's conception of reflective equilibrium:

> The components of a system in reflective equilibrium must be reasonable in light of one another, and the system as a whole *reasonable in light of our initially tenable commitments*. (Elgin 1996, 107, italics added by T.R.)

The constraint that commitments should respect initial commitments is balanced on the other side by the demand that we are searching for a *systematic* account of the subject matter, i.e., principles or a theory that can be used to make justified moral decisions, or to predict events, etc. For this reason, the system should do justice to theoretical virtues like simplicity, fruitfulness, scope, or accuracy (Brun 2020, 25).

Even though this conception of RE is well elaborated (see Chap. 2 for a fuller discussion of it), it might still leave the practitioner who wants to apply the method at a loss: I know that I should start from my initial commitments, but how do I get access to them and how do I select them? How can I determine which theoretical virtues are relevant, and how they should be interpreted? What does it mean for the system and the commitments to be *in agreement*, and how should I resolve conflicts between the two?

Recently, Walden (2013) has defended reflective equilibrium exactly on the grounds that it *does not* in itself give a definite answer to these and similar questions.

According to Walden, RE is anti-essentialist in the sense that it denies that we can say much about our epistemic inputs, methods, and standards in advance of inquiry. On these grounds, he defends RE against opponents who attack the method based on specific ways to define it, e.g., who equate commitments with intuitions and attack RE on the grounds that intuitions are not a plausible input. This defense fits well with Elgin's view that our standards and goals are higher-order commitments which can, in principle, be revised as well (see, e.g., Elgin 1996, 99, 104; 2014, 247–248)—but it doesn't seem to do much for the agent who is searching for a method to apply.

That RE does not say something more essential about the nature of inquiry led Foley (1993) to reject the idea of RE as unhelpful:

> It tells you essentially this: take into account all the data that you think to be relevant and then reflect on the data, solving conflicts in the way that you judge best. On the other hand, it does not tell you what kinds of data are relevant, nor does it tell you what is the best way to resolve conflicts among the data. It leaves you to muck about on these questions as best as you can. (Foley 1993, 128)

However, such worries have not deterred researchers from invoking the method of RE for their work; and Walden acknowledges that for specific projects, the method of RE can be given a more definite character (Walden 2013, Fn. 15). But, as the next section shows, while we can gain valuable insights from previous applications, they fall short of providing the basis for a systematic assessment of how RE can be applied as a method.

1.1.2 Applications of Reflective Equilibrium

Given the claim that reflective equilibrium is an important—maybe even *the* method—for justification, the lack of explicit, well-documented applications is surprising. Beauchamp and Childress (2013) note several unresolved problems about the method of RE, one of them being that there is a striking discrepancy between *claims* to be using the method versus *actual* uses of RE:

> First, ambiguity often surrounds the precise aim of the method. It might be used in reflecting on communal policies, constructing a moral philosophy, or strengthening an individual's set of moral beliefs. The focus might be on judgments, on policies, on cases, or on finding moral truth. Second, it is not entirely clear how to know when our effort to achieve reflective equilibrium is going well, or how to know when we have succeeded. *Explicit uses of the method (by contrast to claims to be using it) are difficult to find in the ethics literature.* Most discussions are heavily theoretical and distant from contexts of practice. [...] Third, the wide-ranging objectives of even a weak wide reflective equilibrium are at minimum intimidating and may be unattainable ideals of both comprehensiveness and coherence. (Beauchamp and Childress 2013, 410, italics added by T.R.)

A look at the literature confirms this assessment: very different purposes are associated with applying RE, and there is a lack of explicit applications. I will say more on these two points in the following. The third point, whether or not the method is too demanding to be applicable, cannot be assessed until the other two points have been addressed.

Aims of Applying RE The objectives that people pursue when stating that they are applying RE differ widely, but can be roughly grouped under four main categories.

The first group sees RE as a method for the justification of (often normative) conceptions, principles, or theories. The most prominent example is probably Rawls (1971), who aims at a theory of justice. Another example is Swanton (1992), who sees the purpose of RE in the justification of a value-conception, in her case that of "freedom". Similarly, De-Shalit and Wolff (2007) want to use RE to give an account of disadvantage that, as part of an egalitarian theory, can be applied to actual society in order to diminish disadvantage. And (Van Thiel and Van Delden 2001) aim to use RE to give an account of the principle of respect for autonomy in health care.

In the second group, RE is seen as an approach to decision-making with respect to a particular (moral) problem. For example, Kushner et al. (1991) refer to it as a "systematic approach to the process of working through an ethical dilemma". In this understanding, RE is a method of decision making that can be used to adjust existing principles and judgments with respect to a particular problem in order to come to a justified decision—in the sense of seeking an answer to the question "What is morally right to do in a particular case?" (e.g., Rutgers 1998, who discusses this for the context of veterinary ethics).

The third group sees RE as a framework for structuring discussions and interpersonal decision-making. For example, Doorn (2010a) sees RE as a framework that can enable decision-making in a pluralist context with different stakeholders, and Schroten (1998) proposes it as a framework for discussions in ethic committees. Similarly, Brandstedt and Brännmark (2020) suggest a version of RE that is intended as "a tool for public reasoning about practical problems which aims to facilitate shared solutions".

Another idea is to use RE to analyze and reconstruct debates, which for example Hahn (2000) demonstrates with respect to the foundational crisis of mathematics. Brun (2014) suggests that RE can be used for modelling and justifying the various steps involved in argument reconstructions.

The different ideas about how RE can be used have even prompted some to speak of different types of reflective equilibrium *methods* (Van der Burg and Van Willigenburg 1998a, 12). This diversity also extends to the structure of the method, i.e., what people think is required in order to apply RE, and is connected with different ideas about the agent(s) involved in the process (for an overview of different ways to conceptualize the social aspects of RE, see Baderin 2017).

In this book, I will investigate RE as a method that aims at justifying principles, theories, or similar. This clearly fits best with the goal of justifying a precautionary principle. Additionally, I will conceptualize it as a method that can be applied by a single epistemic agent, and not, e.g., as a deliberative group project. Both decisions will give the case study a more stringent focus while laying the groundwork for further work on different ways to spell out RE as a method.

How Has RE Been Applied so Far? Consistent with the observation made by Beauchamp and Childress (2013, 410), I could find no example where the actual process of reflective equilibrium, i.e., systematically adjusting commitments and system with respect to each other (as well as with respect to relevant background theories and theoretical virtues), has been done in a way that is both explicit and comprehensive. Either the process is not well-documented—we might grant it to Rawls (1971) that he applied RE, but his arguments are not documented in a way that allows for a clear identification of the steps and elements of the RE process and how they were adjusted under the influence of which considerations (the same holds for De-Shalit and Wolff 2007). Or the process has only been rudimentary started and has not been followed through. Or it is missing altogether, and the application stops after describing part of the input such as a set of commitments or some candidates for background theories. Especially in more empirically oriented studies, the focus is typically on how to obtain inputs for the RE process: for example, by sending out questionnaires (Van Thiel and Van Delden 2001), through conducting semi-structured interviews and analyzing those later (Ebbesen and Pedersen 2007), or by forming one's commitments based on empirical information about the moral experience of others (de Vries and van Leeuwen 2010).

In the cases where the application does not stop after describing (parts of) the setup, typically all of the (few) adjustments take place on the side of the commitments, without confronting them with a systematic account in the form of principles or a theory. One of the few exceptions are Van Thiel and Van Delden (2001), who confront several candidate principles with commitments of practitioners that were identified through questionnaires. They reject those candidates as inadequate and introduce their own proposal, but this new candidate does not again get confronted with the commitments, and remains only a rough sketch.

The studies that made use of structured discussions describe some adjustments of commitments (Doorn 2010b; Van de Poel and Zwart 2009), but these adjustments seem rather to be about learning effects and how participants of a discussion adjust their position under influence of additional input. They thus cannot clearly be attributed to an application of reflective equilibrium.

We also find some demonstrative applications of the method by authors who discuss and develop an account of RE (DePaul 2011; Sinnott-Armstrong 2006), but these are typically very restricted toy examples. For example, Sinnott-Armstrong (2006) demonstrates how narrow RE might work by presenting us with a number of cases and judgments corresponding to them, e.g., "The Watching Case: A baby crawls into a pool of water. A bystander can save the baby's life easily at little cost. Otherwise the baby will drown" and "The Watching Judgment: It is morally wrong for the bystander to let the baby drown", and describes how an agent might generalize moral principles from such judgments, that then get confronted with other cases where they run into problems, e.g. the "Doctrine of Doing and Allowing" runs into problems when confronted with the "Trolley case" and "Trolley judgment" (that it is not morally wrong to pull the lever and direct a runaway trolley from five onto one). Even though adjustments are shown, and even though Sinnott-Armstrong makes clear that it is also possible to adjust commitments, the "equilibration" rather seems to proceed as some sort of "counterexample-philosophy". There is no search for a system of principles that fits with a broad set of initial commitments, and instead the adjustments proceed by searching again and again for counterexamples which are invariably treated as a reason to alter the set of principles. Also, while Sinnott-Armstrong continues to discuss additional conditions and constraints that should be added to RE in order to achieve "wide coherence", e.g., second-order beliefs about the reliability of first-order beliefs, bringing in other people's beliefs, meta-ethical beliefs, metaphysical beliefs, etc., he does not bring this to bear in an (exemplary) application.

While systematic descriptions of an RE process are therefore absent, for descriptions and assessments of RE states matters are indeed even worse. There are some examples in which it is compared how participants of discussions adjusted their positions (Doorn 2010b; Van de Poel and Zwart 2009), but only Doorn (2010b) tries to assess the coherence of judgments, principles, and background at least informally. Van Thiel and Van Delden (2001) do compare theory candidates with commitments, but not systematically, and although they do make a new theory proposal, they do not assess how coherent the position would be with that system, and what adjustments might still be necessary. Explicit criteria for assessing positions and identifying weaknesses are thus still an important desideratum for applications of RE. This and other desiderata are listed in the next section, where I summarize the most pressing challenges for applying reflective equilibrium as a method.

1.1.3 Desiderata for a Case Study

As the above section shows, previous applications of RE fail to sufficiently exemplify how the method can be applied. This is not to discredit their academic merit. Most of them were concerned with concrete and specific challenges and problems, and merely wanted to use RE as a method to tackle these challenges. However, the deficiencies of existing applications make it difficult to assess what role RE can play as a method. This makes a detailed case study necessary in order to test whether it is possible to find instances of its elements in actual applications, and whether its criteria can be used in insightful ways to guide a process of adjustments, and to assess whether a position is justified.

In particular, we can identify the following challenges for applying reflective equilibrium as a method:

- The aims and purpose of applying the method are unclear;
- It is unclear what kinds of input, and whose inputs, to select—are the relevant commitments intuitions, (considered) judgments, or beliefs? Should they be of philosophers, practitioners, or the broader public?;
- It is unclear what it entails to apply the method, i.e., what its steps are and how to complete them, and how to decide about adjustments;
- It is unclear how to measure progress and to assess whether or not one did actually reach a state of reflective equilibrium;

- On the one hand, there are worries that, because of its coherentist and holistic character, the method might be too demanding to be actually applicable;
- On the other hand, there are worries that, because it has no fixed standards, RE is too vacuous to provide any useful guidelines or real constraints, and thus falls short of being a method.

Clearly, the last two worries—whether RE is either too demanding or too vacuous to be applied as a method—cannot be assessed until the other problems have been solved and a real attempt has been made to apply RE. To provide a real test for RE as a method, this application should be to an actual, complex problem, instead of working with simplified toy examples. Thus, we can identify the following **desiderata for a case study on the applicability of RE as a method**:

1. Reflective equilibrium (RE) should be spelled out as explicitly and precisely as possible, including: (i) its theoretical and methodological foundations, (ii) the aim of the method, and (iii) how RE can be specified for particular applications.
2. RE should be applied to an actual, complex problem in order to provide a real test case.
3. The initial position of the process of adjustments should be explicitly described, e.g., the input and the specified criteria.
4. The application should be traceable step-by-step, describing what was adjusted when and with respect to what, and how the adjustment in question can be defended with the RE criteria.
5. The application should include a detailed comparison of the initial and resulting position, including assessing the resulting position with respect to RE criteria and whether or not a state of RE was reached.
6. The application and its results should be evaluated and critically discussed in order to learn from it for the use of RE as a method.

A case study that aims to meet these desiderata will allow us to assess the benefits and challenges of applying RE. If RE can be successfully applied, the case study will also help to provide guidelines for further applications. For example, it is an open question whether the process of adjustments that is part of RE should be understood as a method that one can follow step-by-step when trying to justify something. Is reflective equilibrium best understood as a constructive method, i.e., a method that describes a procedure that one can follow in order to make epistemic progress? Or is it better understood as spelling out requirements for justification that we can use for the reconstructive appraisal of epistemic positions? By testing RE in a constructive application with the goal of formulating and justifying a precautionary principle, this book aims to provide important insights on these questions.

In the next section, I introduce the topic of precautionary principles, which was selected as the topic of the case study in order to meet desideratum (2). Section 1.3 then gives an overview about the whole book, describing how and where the desiderata (1)–(6) are addressed.

1.2 Precautionary Principles as a Test Case for Applying Reflective Equilibrium

The question of how a precautionary principle (PP) can be justified is an ideal test case for the applicability of RE. It provides an actual, complex problem: the basic idea behind PPs is that we have to take action to prevent harm even if we are uncertain about its likelihood or extent. It is often summarized as "better safe than sorry". However, it is controversial how this general idea of precaution should be spelled out as a principle.

PPs emerged first in regulatory contexts, arguably as an answer to ineffectual policies of the past which failed to protect the environment and/or human health in important respects (Harremoës et al. 2001). While talk of "the" PP in the singular is common, formulations and interpretations vary so greatly that, if only to be on the safe side, it is appropriate to use the plural form and speak of precautionary principles (cf. Hartzell-Nichols 2013).

Precautionary principles are often seen as a supplement or even as an alternative to traditional approaches to risk regulation, like quantitative risk assessment and cost-benefit analysis. Because the latter require knowledge about the relationship between options and outcomes as well as the probabilities of the various possible outcomes, they condemn us to idleness in cases where we lack that sort of knowledge, if we consider them the only maintainable approach to risk management. Thereby, so the criticism goes, they inhibit immediate measures which would be necessary in order to avoid serious damage. PPs, in contrast, demand action even if this knowledge is not or only partially available.

However, precautionary principles also have been criticized for various reasons, such as being too vague to be action-guiding, paralyzing the decision-process through conflicting recommendations, or being anti-scientific and promoting a culture of fear (for an overview of the criticisms see, e.g., Randall 2011; Sandin et al. 2002).

The challenge of how to spell out the basic idea of precaution in the form of a systematic precautionary principle is thus a suitable test case for the applicability of RE: first, it presents us with a practical problem in which we, as imperfect epistemic agents, have to find a justified principle as a basis for our decisions. Second, there are already a number of existing proposals for PPs to draw on. Third, PPs are often confronted with objections like that they lead to incoherent recommendations. Resolving such inconsistencies seems exactly what RE is designed for. Additionally, it is a topic of practical relevance in the face of, e.g., current threats arising from global climate change.

1.3 Overview of this Book

This book tests a specific conception of reflective equilibrium (RE) in a case study on an actual, complex problem: how can a precautionary principle (PP) be justified? In doing so, it addresses the desiderata (1)–(6) identified in Sect. 1.1.3.

To clarify the tasks ahead, it is useful to make the distinction between epistemology, methodology, and method. Although RE is often referred to as either a method or a methodology, it is rarely explained what is meant by these terms—and sometimes they seem to be used almost interchangeably.[3] To further complicate things, RE is also often called an epistemology, or account of justification (e.g., Tersman 1993). Thus, in order to obtain a clearer picture of what it means to apply RE *as a method*, it is important to distinguish these three categories.

For the purpose of this book, I propose the following distinctions (cf. Ackerly and True 2013): firstly, **methods** are concrete tools and techniques of research, in the sense of a set of instructions or a specification of steps which should be followed to achieve a given (sub-)objective (see also Caws 1967, 339; McPherson 2015, 653). Secondly, **methodology** is a theory and analysis of how research should proceed. That is, it describes the general research strategy by answering questions such as which method, or combination of methods, is adequate for pursuing a specific research question. As Ackerly and True (2013, 137) put it, a methodology does not prescribe particular methods no matter the question, but is better understood as a framework which guides decisions at various stages during the research process. And thirdly, **epistemology** is concerned with a theory and analysis of what has epistemic value, e.g., under what conditions a belief qualifies as knowledge.

Consequently, if we want to apply RE as a method for the justification of a principle such as a precautionary principle, we need, firstly, to be clear about its epistemological foundations, e.g., what it means for a principle to be justified. Secondly, we need to develop a methodology that guides the specification and application of the method with respect to a specific objective, and in light of the underlying epistemology.

Chapter 2 describes the epistemological foundation of RE as a method of justification. It describes the conditions under which an epistemic position is justified through being in a state of reflective equilibrium. The chapter introduces the different elements that are part of RE—commitments, a system, background elements, and theoretical virtues—as well as the relations between them. As explained above in Sect. 1.1.1, I adopt a conception of RE from the works of Elgin, Brun, and Baumberger. While this conception is well elaborated, it is concerned primarily with the epistemological analysis of RE as an account of justification. As

[3] For example, Kelly and McGrath (2010, 352; italics added by T.R.) write that what makes the *method* appealing to many philosophers is that it seems to provide a relatively down-to-earth *epistemology*; and that "the *method* has generally been most popular as an account of the *methodology* [for certain domains]." One of the few exceptions is McPherson (2015), who explicitly introduces his use of "method" and "methodology".

such, it does not tell us how the elements and criteria of RE should be specified for a particular project. For example, it tells us that we need a system, like a theory or a set of principles, and that this system should do justice to theoretical virtues. But questions such as which theoretical virtues are relevant for the development of a particular system is not inherent to RE as a general account of justification.

We thus need to develop a methodology that allows us to spell out RE as a specific method for a specific justificatory project, which is the main task of Chap. 3. This chapter develops guidelines for specifying RE as a method, and discusses methodological issues that need to be resolved. For the purpose of the present book, this is done with respect to two important restrictions, which will help to further sharpen the method: firstly, the method should be applicable by a single epistemic agent, and not require group deliberation or similar group processes. Secondly, I am concerned with reflective equilibrium as a method that aims at justifying principles, theories, or similar; and not, for example, as a method for finding a justified consensus, or as a decision framework. Chapters 2 and 3 thus contribute to desideratum (1), to be as explicit and precise as possible about the theoretical and methodological foundations of RE, and to show how RE can be specified as a method with respect to a particular aim.

Chapter 4 gives a survey about different interpretations of precautionary principles, proposed justifications for them, as well as objections and possible rejoinders. This contributes to desideratum (2), that RE should be applied to an actual, complex problem. The results from Chap. 4 also provide the basis to identify the input of the RE process, i.e., candidates for commitments, principles, and background elements of the RE application. This input, and the reasons for selecting it, is described in Chap. 5, which addresses desideratum (3), an explicit description of the initial position from which the process of adjustments starts. As part of Chap. 5, the method of RE is also specified with respect to the particular project of justifying a precautionary principle, which further contributes to desideratum (1).

Chapters 6–8 then present the actual case study. To cover as many aspects of applying RE as possible, the application is divided into three parts: Chap. 6 demonstrates how RE can guide systematization and theory development, Chap. 7 describes step-by-step how commitments and principles are adjusted in alternation, and Chap. 8 shows how a specific position can be spelled out and appraised with the RE criteria. Chapters 6–8 thus address desideratum (4), a traceable step-by-step application that makes it clear when adjustments are made and how they are defended. Chapter 8, in which the equilibration process comes to a preliminary end point, also addresses desideratum (5) by assessing the resulting position and discussing whether a state of reflective equilibrium was reached. With respect to precautionary principles, the application yields a rights-based precautionary principle.

Intermediate results are discussed after each chapter of the application. Chapter 9 summarizes them and discusses what we can learn from the case study for the applicability of reflective equilibrium as a method. It thereby addresses desideratum (6). As the case study shows, RE can be specified in a way that makes it applicable while being neither too demanding nor too permissive. In the case study, RE does put real constraints on the justification process, while allowing us to structure the process of adjustments and facilitating the search for further relevant considerations. However, the application in this book is also very complex, partly due to the fact that, as a case study, it has a strong focus on how RE can be applied. Not everyone might want to engage with the method at this level of detail. Thus, Chap. 9 also discusses how one can make use of the methodological benefits from RE without having to spell out every aspect in as much detail as is done in Chaps. 6–8.

References

Ackerly B, True J (2013) Methods and methodologies. In: Waylen G, Celis K, Kantola J, S Laurel W (eds) The Oxford Handbook of Gender and Politics, pp 135–153

Baderin A (2017) Reflective equilibrium: individual or public? Soc Theory Pract 43(1):1–28

Baumberger C, Brun G (2017) Dimensions of objectual understanding. In: Grimm SR, Baumberger C, Ammon S (eds) Explaining understanding: new perspectives from epistemology and philosophy of science. Routledge, New York, pp 165–189

Baumberger C, Brun G (2021) Reflective equilibrium and understanding. Synthese 198(8):7923–7947. https://doi.org/10/ggkp4w

Beauchamp TL, Childress JF (2013) Principles of biomedical ethics, 7th edn. Oxford University Press, Oxford

Brandstedt E, Brännmark J (2020) Rawlsian constructivism: a practical guide to reflective equilibrium. J Ethics 24(3), 355–373. https://doi.org/10/ggvr8n

Brun G (2013) Reflective equilibrium without intuitions? Ethical Theory Moral Pract 17(2):237–252. https://doi.org/10.1007/s10677-013-9432-5

Brun G (2014) Reconstructing arguments: formalization and reflective equilibrium. History of Philosophy and Logical Analysis 17(1):94–129

Brun G (2016) Explication as a method of conceptual re-engineering. Erkenntnis 81(6):1211–1241. https://doi.org/10.1007/s10670-015-9791-5

Brun G (2020) Conceptual re-engineering: from explication to reflective equilibrium. Synthese 197(3):925–954. https://doi.org/10.1007/s11229-017-1596-4

Caws P (1967) Scientific method. Encycl. Philos. 7:339–343

Daniels N (1979) Wide reflective equilibrium and theory acceptance in ethics. J. Philos. 76(5):256–282

De-Shalit A, Wolff J (2007) Disadvantage. Oxford University Press, Oxford

de Vries M, van Leeuwen E (2010) Reflective equilibrium and empirical data: third person moral experiences in empirical medical ethics. Bioethics 24(9):490–498. https://doi.org/10.1111/j.1467-8519.2009.01721.x

DePaul M (2011) Methodological issues: reflective equilibrium. In: Miller C (ed) Continuum companion to ethics, continuum

DePaul MR (1993) Balance and refinement: beyond coherence methods of moral inquiry. Routledge, London

Doorn N (2010a) Applying Rawlsian approaches to resolve ethical issues: inventory and setting of a research agenda. J Bus Ethics 91(1):127–143

Doorn N (2010b) A Rawlsian approach to distribute responsibilities in networks. Sci Eng Ethics 16(2):221–249

Ebbesen M, Pedersen BD (2007) Empirical investigation of the ethical reasoning of physicians and molecular biologists—the importance of the four principles of biomedical ethics. Philos Ethics Humanit Med 2(1):23

Elgin CZ (1996) Considered Judgment. Princeton University Press, Princeton

Elgin CZ (2014) Non-foundationalist epistemology: holism, coherence, and tenability. In: Steup M, Turri J, Sosa E (eds) Contemporary debates in epistemology, 2nd edn, Contemporary debates in philosophy. Wiley Blackwell, Chichester, pp 244–255
Elgin CZ (2017) True enough. The MIT Press, Cambridge
Foley R (1993) Working without a net: a study of egocentric epistemology. Oxford University Press, New York
Goodman N (1951) The structure of appearance. Harvard University Press, Cambridge
Goodman N (1978) Ways of worldmaking. Hackett Publishing, Indianapolis
Goodman N (1983) Fact, Fiction, and Forecast, 4th edn. Harvard University Press, Cambridge
Hahn S (2000) Überlegungsgleichgewicht(e): Prüfung Einer Rechtfertigungsmetapher. Alber
Harremoës P, Gee D, MacGarvin M, Stirling A, Keys J, Wynne B, Vaz SG (eds) (2001) Late lessons from early warnings: the precautionary principle 1896–2000. Office for Official Publications of the European Communities, Luxembourg
Hartzell-Nichols L (2013) From 'The' Precautionary Principle to Precautionary Principles. Ethics, Policy and Environment 16(3):308–320
Kelly T, McGrath S (2010) Is reflective equilibrium enough? Philos Perspect 24(1):325–359. https://doi.org/10/bmh3g5
Kushner T, Belliotti RA, Buckner D (1991) Toward a methodology for moral decision making in medicine. Theor Med 12(4):281–293. https://doi.org/10.1007/BF00489889
Lewis D (1983) Philosophical papers, vol I. Oxford University Press, Oxford
McPherson T (2015) The methodological irrelevance of reflective equilibrium. In: Daly C (ed) The Palgrave Handbook of Philosophical Methods. Palgrave Macmillan, New York, pp 652–674
Randall A (2011) Risk and precaution. Cambridge University Press, New York
Rawls J (1971) A theory of justice. Belknap Press, Cambridge
Rawls J (1974) The independence of moral theory. Proceedings and Addresses of the American Philosophical Association 48:5–22. https://doi.org/10.2307/3129858
Rawls J (2001) Justice as fairness: a restatement. Harvard University Press, Belknap
Rutgers B (1998) The use of the reflective equilibrium method in normative veterinary ethics. In: van der Burg W, van Willigenburg T (eds) Reflective equilibrium: essays in honour of Robert Heeger, library of ethics and applied philosophy, vol 2, Springer, Netherlands, pp 231–237
Sandin P, Peterson M, Hansson SO, Rudén C, Juthe A (2002) Five charges against the precautionary principle. J Risk Res 5(4):287–299
Scanlon TM (2003) Rawls on justification. In: Freeman S (ed) The Cambridge companion to Rawls, Cambridge companions to philosophy. Cambridge University Press, Cambridge, pp 139–165
Scheffler I (1954) On justification and commitment. J. Philos. 51(6):180. https://doi.org/10.2307/2021776
Schroten E (1998) The 'Herman Case': the usefulness of the wide reflective equilibrium model for ethics committees. In: van Willigenburg T, van der Burg W (eds) Reflective equilibrium: essays in honour of Robert Heeger, library of ethics and applied philosophy, vol 2, Springer, Netherlands, pp 219–229
Sinnott-Armstrong W (2006) Moral Skepticism. Oxford University Press, Oxford
Swanton C (1992) Freedom: a coherence theory. Hackett Publishing, Indianapolis
Tersman F (1993) Reflective equilibrium: an essay in moral epistemology. Lagerblads tryckeri AB, Karlshamn
Van de Poel I, Zwart SD (2009) Reflective equilibrium in R & D networks. Sci. Technol. Hum. Values 35(2):174–199. https://doi.org/10.1177/0162243909340272
Van der Burg W, Van Willigenburg T (1998a) Introduction. In: Van der Burg W, Van Willigenburg T (eds) Reflective equilibrium: essays in honour of Robert Heeger, library of ethics and applied philosophy, vol 2, Springer, Netherlands, pp 1–25
Van der Burg W, Van Willigenburg T (eds) (1998b) Reflective equilibrium: essays in honour of Robert Heeger, library of ethics and applied philosophy, vol 2. Springer, Netherlands
Van Thiel GJ, Van Delden JJM (2001) The principle of respect for autonomy in the care of nursing home residents. Nurs Ethics 8(5):419–431. https://doi.org/10.1177/096973300100800506

Van Thiel GJ, Van Delden JJM (2010) Reflective equilibrium as a normative empirical model. Ethical perspectives-Katholieke Universiteit Leuven 17(2):183–202

Walden K (2013) In defense of reflective equilibrium. Philos Stud 166(2):243–256. https://doi.org/10.1007/s11098-012-0025-2

Chapter 2
Theoretical Foundations of Reflective Equilibrium

Before we can address the question of how to apply reflective equilibrium (RE) as a method, we have to be clear about how we understand RE and what we take its elements and criteria to be. Only then can we test the implications of this conception for its application as a method. This is expressed in desideratum (1), which was identified in Chap. 1:

Desideratum 1 Reflective equilibrium should be spelled out as explicitly and precisely as possible, including: (i) its theoretical and methodological foundations, (ii) the aim of the method, and (iii) how RE can be specified for particular applications.

The present chapter describes the theoretical foundations of a specific conception of RE, that is, its elements and criteria. It thereby addresses part (i) of this desideratum. Parts (ii) and (iii) of the desideratum are addressed in Chap. 3, where I discuss how we can obtain a method from this theoretical conception.

I start by giving an overview of the main ideas of RE in Sect. 2.1, before focusing on the elements of RE and the relations between them. The relation of *Agreement*, i.e., the sought-after balance between commitments and a systematic account of these commitments, is addressed in Sect. 2.2. This relation is central for RE, and is predominantly associated with it. However, in the conception adopted here, RE does not reduce to coherence between commitments and a system: we do not only start from our commitments about the subject matter in question, but also have to respect them throughout the process. What this means is discussed in Sect. 2.3. Section 2.4 addresses the role of theoretical virtues for the system, while Sect. 2.5 explains what makes RE "wide", that is, how the justificatory project in the foreground is situated in a broader context of background theories, background information, and background assumptions. Section 2.6 sums up the criteria of RE, and sketches the way ahead towards the case study.

T. Rechnitzer, *Applying Reflective Equilibrium*, Logic, Argumentation & Reasoning 27, https://doi.org/10.1007/978-3-031-04333-8_2

2.1 The Idea of Reflective Equilibrium

Very roughly, the core idea of (so-called "wide") RE is that we start from our existing judgments about relevant cases and search for fitting systematic principles, which, in turn, can be applied to new cases. When conflicts arise between judgments and principles, both sides are adjusted mutually in a process guided by epistemic goals and supported by accepted background theories. When a coherent state—an equilibrium—is reached as a result of this process, we can consider both our judgments and our principles to be justified.

In order to obtain a specific RE conception that can be tested in a case study, this rough idea needs to be spelled out. To do this, I draw on the most elaborate and fleshed-out conceptions that can be found in the literature, in particular on the works of Elgin (1996, 2017), Brun (2013, 2016, 2020), and Baumberger and Brun (2017, 2021). As we will see, this RE conception is more complex than just bringing particular judgments and general principles into coherence with each other: it requires an epistemic position to meet six criteria in order to be in a state of reflective equilibrium. In this section, I introduce the main ideas of this conception of reflective equilibrium, before spelling them out in more detail in the next sections.

Commitments vs. System Standard accounts of RE typically see the difference between judgments and principles in terms of particular vs. general, but already Rawls (1974, 289) pointed out that judgments can also be general. Hence, Brun (2013, 240) argues that the main difference between principles and judgments has nothing to do with their content or their form, but that it is their function that is different. He argues that principles are *part of a system*, while judgments involve a certain degree of *commitment*—as minimal as it may be. This means that we can have parts of a system that are more or less a restatement of a commitment (cf. Knight 2017, 51–52), i.e., we can be committed to the judgment "one should not lie" while also using the principle "one should not lie" to systematically account for commitments. Thus, I will usually talk of *commitments* and *(parts of) a system*, where a system can be, e.g., a theory, a set of principles, or a model. Of course we can be committed to particular judgments, and general principles can be part of a system, but the relevant distinction is between the attitude of commitment on the one side, and the ability to provide a systematic account on the other. Together, the set of commitments and the system form a *position*; this is the object of the justification via RE. I say more on commitments, systems, and the relation between them in Sect. 2.2.

Agreement Typically, the idea of reflective equilibrium is associated with bringing commitments and systems into agreement through mutual adjustments. This relation of *agreement* is normally understood as *coherence*, raising the question what coherence is, precisely. It is commonly agreed that coherence at the minimum requires consistency as a necessary condition, but also something more (Van Thiel and Van Delden 2009, 236). In the RE context, I take agreement between commitments and system to require at least that commitments and system are consistent, and that the commitments can be inferred from the system—although not necessarily in a strict deductive sense, and given relevant background information (Brun 2013, 241).

Given how central the search for agreement between systems and commitment is for RE, it is not surprising that it is often characterized as a *coherentist* account of justification: elements of an epistemic position are justified through being part of a coherent position, i.e., through their relations to other elements (Cath 2016, 216).

The RE Criteria: More than Coherence However, the idea of RE involves more than coherence (see Baumberger and Brun 2021, 7931, as well as the references given there). As Brun (2020, 948–50) argues, there is a second reading of the "reflective equilibrium"-metaphor: namely, that the criterion (1) of agreement between commitments and system has to be established against the two further demands that (2) the resulting system has to do justice to theoretical virtues, and that (3) the resulting commitments have to (a) respect input commitments, and (b) at least some of them should have credibility that is independent of the current RE state (see also Baumberger and Brun 2021, 7931–35). The demands (2) and (3) are "pulling" from two sides on the relation between commitments and system, putting constraints on how the agreement can be reached: that the system has to be *systematic* in the sense of having theoretical virtues like simplicity or comprehensiveness blocks the conservative strategy of establishing consistency via the path of least resistance, i.e., avoiding adjustments of commitments whenever possible. And that resulting commitments have to have some independent credibility and also to respect *input commitments*, i.e., the commitments we have independently of the RE process, means that we cannot just formulate a very simple and comprehensive system, accepting everything that follows from it at the expense of rejecting all commitments that conflict with it.

Additionally, there is the demand that (4) the position should not only be internally but also externally coherent, meaning that the resulting system should be supported by *background theories* (Daniels 1979).

Weak Foundationalism Because of criterion (3)—that commitments have to respect input commitments, and should have some credibility independent of the current RE state—I understand RE as a weakly foundationalist theory of justification (Brun 2013; Elgin 2014; Hansson 2007). This means that justification is not only derived from coherence alone: instead, a set of initially held commitments that have a minimal degree of credibility is revised in order to enhance their credibility by fitting them into a coherent account, which can also mean *rejecting* some of these initially credible commitments. Thus, justification via reflective equilibrium is not purely coherentist, because it requires that at least some of the resulting commitments have a degree of credibility that is independent of the coherence of the resulting position (Baumberger and Brun 2021, 7932). But it is only weakly foundationalist (BonJour 1985, 28–29), since a commitment can never be justified by independent credibility alone: for this, the commitment needs to be part of a coherent position that is in a state of reflective equilibrium. I say more on independent credibility below in Sect. 2.3.

Pragmatic-Epistemic Goals Reaching a state of reflective equilibrium is a matter of degree—commitments and a system can be more or less in agreement, commitments can have more or less independent credibility, the system can exhibit virtues like simplicity or scope to a higher or a lower degree, and so on.[1] Because of the plurality of goals that a target position in RE should meet, there is no reason to think that all of them can be maximized at the same time: trade-offs are usually unavoidable.

> [Doing] justice to a plurality of epistemic goals can involve trade-offs between any of them. Increasing the simplicity of a theory may only be feasible by discarding some independently credible commitments and thereby rejecting pieces of evidence. On the other hand, maintaining credibility blocks oversimplifications and sweeping generalizations one may be tempted to accept in the name of systematicity. That trade-offs are typically unavoidable is also a reason why we speak of "doing justice to" rather than "realizing" epistemic goals. Although justification calls for taking epistemic goals seriously, it would be unrealistic to insist on theories which effectively reach all those goals simultaneously. (Baumberger and Brun 2017, 178)

Which configuration of goals is relevant, and how much weight they should have, depends both on the subject matter and the specific pragmatic-epistemic objective that is pursued in the project of justification (Elgin 1996, 105; Baumberger and Brun 2017, 178; 2021, 7928). For example, for Rawls' project with the pragmatic-epistemic objective of justifying principles of justice for a liberal society, Rawls (1999, 113–17) sees it as important that the principles should be general, universal, publicly acknowledged, impose an ordering on conflicting claims, and be accepted as final instances. Kuhn (1977, 322) names "accuracy, consistency, scope, simplicity, and fruitfulness" as some of the standard criteria for the evaluation of the adequacy of a theory, but depending on the specific objective, their importance will vary. For example, if we want to develop a regional weather model that makes exact predictions for mountain valleys, precision will be more important than if we want to develop a climate model for the purpose of understanding the basic mechanism of global climate change. In the latter case, however, simplicity and scope might be more important.

Justification and Pluralism in RE Figure 2.1 gives a schematic overview of the main elements and requirements of RE. It shows how the position in the foreground is developed through adjusting commitments and a system both with respect to each other as well as with respect to the other constraints on a position in reflective equilibrium. We can thus distinguish between the process of searching for reflective equilibrium, and a state of being in reflective equilibrium. It is worth pointing out that, firstly, the RE criteria can be met to different degrees. This means that, secondly, it could be that there are several possible positions that, on balance, meet a configuration of these criteria equally well. There will often be different ways to resolve trade-offs between, e.g., theoretical virtues in the sense that there is no

[1] But note that this is not the case for consistency, which does not admit of degrees and is a minimal, necessary condition for a position to be in a state of reflective equilibrium.

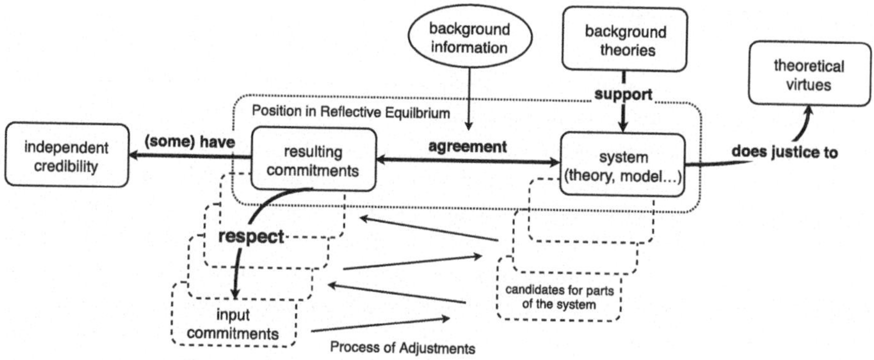

Fig. 2.1 A schematic overview of reflective equilibrium

candidate system that is overall more virtuous than all other candidates, or between theoretical virtues of a system versus its ability to account for current commitments. This means that there can be a plurality of RE states that might be reached in different ways. But as long as they are equally plausible, i.e., overall meet the RE criteria to the same degree, choosing between them is a question of practicality, not of being justified. Equally plausible positions are justified to the same degree—but should of course be given up in favor of better justified positions (Elgin 1996, 107; 119; 2017, 66; 87–88). Thus, a fifth criterion for a position to be in a state of RE is (5) that it is at least as plausible with respect to the RE criteria as relevant alternatives.

There is no guarantee that an RE state will be reached. RE gives us standards of justification, but there is no guarantee that by conducting a process of adjustments, we will automatically end up with a justified position. The justification RE provides is inherently provisional, and we always have to continue to assess our positions in order to review whether we can still reasonably claim to be justified (Elgin 1996, 12).

In the following sections, I address each of the RE criteria in turn, before summing them up and sketching the way ahead in Sect. 2.6.

2.2 Agreement between Commitments and System

The relation between commitments and a system is at the center of reflective equilibrium: we aim to arrive at justified commitments concerning a subject matter via supporting them with a system—e.g., one or more principle(s), a theory, or a model—where this system is, in turn, justified through being brought into agreement with our commitments. In the following, I elaborate in turn on commitments, the system, and the relation of agreement.

2.2.1 Commitments

Given a certain subject matter—e.g., fairness—we have various commitments about how actions and outcomes should be evaluated, or how we should or should not act. These can be explicit, i.e., statements which we openly endorse, or implicit, e.g., commitments that get expressed in our actions, or that we would endorse if they were presented to us. For example, as an only child, I might never have considered whether it is unfair if always only one of two siblings gets gifts, but will assert it as soon as presented with the situation. The status of being a commitment comes from being held, as Elgin (1996, 160) puts it, not from being aware that one holds it.

We can be committed to intuitions (Brun 2013), beliefs, judgments, or (what we take to be) considered judgments, but also to non-propositional content like pictorial representations or emotions (Elgin 1996, chapter V). Commitment comes in different degrees, and it can also be "minimal or feeble" (Brun 2013, 240), as long as it is considered to have something speaking favor of it or to be at least a starting point at which we find ourselves (Baumberger and Brun 2021, 7930). We can be committed to a broad range of propositions, both particular and general (Elgin 1996, 102)—the difference between commitments about a subject matter and parts of a system is therefore neither one of content nor of form, but a functional distinction (Baumberger and Brun 2017, 172).

We start from our initial commitments about a subject matter, but to make progress we need to develop a systematic account of said subject matter. And merely listing the commitments we initially hold will not meet this purpose (Elgin 1996, 103; Baumberger and Brun 2021, 7930). Our commitments are apt to be inconsistent with each other; or their relationship might be unclear altogether (Elgin 1996, 106).

For example, if I judge that situation *a* is unfair and situation *b* fair, it might not directly be clear whether or not the situations are similar in a way that makes it inconsistent to hold both these judgments at the same time. Some of what seems like an initially credible commitment might turn out to be based on prejudice. Or there might be inferences between them that we did not consider so far, e.g., a subset of commitments where each one seems plausible on its own, but that might together commit us to something that we actually do not want to accept. In short, our initial commitments do not allow us to meet our pragmatic-epistemic goals, and this is why we start a process of inquiry and justification in the first place.

Merely collecting and writing down our commitments in a list does not enable us to directly infer anything for new cases, as we cannot be clear what exactly the relevant features are that make us classify an act as, e.g., fair or unfair. And neither will it allow us to distinguish between justified commitments and those based on prejudice, bias, or misinformation. To identify and test (candidates for) relevant features, we have to formulate a system that is in agreement with our commitments, i.e., that names relevant features and can account for the commitments. This system is formulated through a process that will also transform our commitments.[2]

We can distinguish between input, current, and resulting commitments, meaning the commitments we have independently of the RE process, the ones we have at a given point during the process, and the ones that result at the end of the process. These distinctions are due to the nature of the RE process, during which some of the commitments that we started out with will be adjusted or rejected, while additional commitments can be accepted. Thus, the input and the resulting commitments might differ significantly.

Among input commitments, the additional distinction between initial and emerging commitments can be made (Baumberger and Brun 2021, 7932). Initial commitments are a sub-class of the input commitments, namely the input commitments the RE process initially starts out with. Emerging commitments are another subclass of the input commitments. They are commitments which are made explicit during the RE process but neither because they are inferred from the system nor as a direct result of an adjustment to the position. This might happen because during the RE process we can be confronted with new considerations that never occurred to us before. Thus, like the initial commitments, those emerging commitments enter the process as input. Emerging commitments need to be distinguished from what Baumberger and Brun (2021, 7932) call "purely inferential" commitments, that is commitments which the epistemic agent accepts purely because she inferred them from the system currently under construction. As we will see in Sect. 2.3, this difference is important for the constraint that resulting commitments should respect input commitments.

2.2.2 System

The set of commitments that we have before we start the RE process expresses our commitments concerning a subject matter. The system is constructed for the purpose of clarifying and systematizing our commitments in order to do justice to the pragmatic-epistemic goals that drive the process of justification. It fulfills a different function than the commitments and expresses a change in perspective on the subject matter, aiming to grasp it from a theoretical point of view. This functional distinction remains true even when a reflective equilibrium is reached and we are also committed to the system in the indirect sense of being committed to everything that can be inferred from the system.

That this distinction is purely functional, and not fixed on content or form, is illustrated by Brun (2020) with the example of *modus ponens*: On the one hand, we

[2] However, the application of RE in itself cannot guarantee that we get rid of, e.g., all our (moral) prejudices. This will sometimes only be possible if we consider epistemic background theories that allow us to identify specific commitments as ill-founded.

can see the inference rule $\phi \rightarrow \psi; \phi \vdash \psi$ as a principle, i.e., a part of a logical system that, e.g., characterizes valid logical inferences. On the other hand, we can be committed to accept all inferences of this form as valid. If an RE was reached, both will be the case.

The system can include elements such as principles, theories, models, or category schemes (for a discussion of category schemes, see Elgin 1996, 136). Parts of the system can be law-like or general, but they do not need to be (see, e.g., Gertken 2014): the important part is that the system is in agreement with the commitments while also doing justice to theoretical virtues.

I understand the system as the organizing element which is constructed in order to identify which features of the commitments are the relevant ones, how the different commitments relate to each other, and to make explicit what further considerations are relevant and should be added. Going back to the fairness case, we could, e.g., compare a system that picks out equality as the relevant feature with one that identifies desert as the main consideration for fairness. Which of them turns out to be more plausible will depend on how well it can account for the current commitments, whether we are willing to commit to the implications of the system for situations we did not consider before, and whether or not we can use this system for the purpose we had in mind. For example, a system that is restricted to fairness in interpersonal relations will most likely not be suitable as an account for fairness in jurisprudence.

2.2.3 Agreement between System and Commitments

The sought-after relation of *agreement* between commitments and system is often identified with coherence. To be in agreement, commitments and system have to be consistent with each other, but this is only a necessary, not a sufficient condition: also completely unrelated propositions can also be consistent with each other. What else is needed on top of consistency to count as agreement will differ depending on the specific project of justification one pursues. One sensible proposal for a minimal characterization of agreement is that (a) the commitments are consistent in themselves, (b) the system is consistent in itself, (c) commitments and system are consistent with each other, and (d) the system accounts for the commitment in the sense that the latter can be inferred from the former (given relevant background information). “Inference” includes, but does not reduce to, deductively valid arguments. Requiring that epistemic coherence always rests on logical inferences would be too demanding. Weaker relations, e.g., “increases the probability that” or “makes it more reasonable to believe that” can form the basis for coherence (Hansson 2006, 100).

Arguably, the agreement criterion is a necessary requirement for a position to be in a full state of reflective equilibrium. There is no guarantee that it can be reached, though: maybe the best position that we can reach still includes conflicts between commitments and a system. Justification in the form of reflective equilibrium is not trivial to reach.

However, agreement is not sufficient for RE, either: agreement could be reached, e.g., by only making the most minimal adjustments to the commitments in order to resolve inconsistencies, and then writing the list of the commitments down as a "system"—which would of course be able to account for the commitments since it would be identical with them. If we can consistently commit to everything that can be inferred from this "system", agreement would have been reached.[3] However, we would not be willing to call this cleaned up list a "system", nor would it meet the goals that motivated us to start revising our commitments. On the other hand, it seems that too-drastic revisions of commitments have to be blocked, too: we should not simply abandon the subject matter and start pursuing another project of inquiry. The latter might seem dubious and carry the danger of making RE too conservative (cf. Dutilh Novaes 2020, n. 5)—could it not, e.g., result in us staying committed to biases and prejudices?

In the next section, I distinguish two senses of how input commitments should be *respected*—and argue that, if understood in the right way, this "respect" does not make RE too conservative, but actually contributes to its justificatory power.

2.3 Respecting Input Commitments and the Criterion of Independent Credibility

There are (at least) three demands on commitments: As we have already seen, (i) commitments should be in agreement with the system, which also requires that they are consistent. Additionally, (ii) they should have *independent credibility*, i.e., some credibility that is independent of the current RE process, and (iii) *resulting commitments* should *respect input commitments* adequately (see Baumberger and Brun 2021; Brun 2020, for arguments in favor of making these distinctions). In this section, I focus on the demands that are particular to the commitments, i.e., independent credibility and respecting input commitments in order not to abandon the subject.

2.3.1 *Independent Credibility*

Reflective equilibrium never starts completely from scratch (Scheffler 1954, 188): we always have some sentences that we hold as acceptable or plausible, even if we are lacking the justification via RE. These are the *input commitments* that we have independently, or before, the RE process. We are committed to these sentences (or

[3] Of course, it is highly dubious that the inferences from a "system" obtained by such a strategy would be consistent.

acts, or values, etc.) because we ascribe some *independent credibility* to them, i.e., credibility they have independently of being justified via RE.

Reasons for ascribing independent credibility to a commitment can be because (a) it accommodates some evidence, which for empirical projects might mean observations or testimony, and for non-empirical projects could, e.g., refer to intuitions (Baumberger and Brun 2017, 177). Another reason for ascribing some minimal degree of independent credibility to a commitment is simply (b) that so far nothing speaks against holding this commitment. Such very minimal commitments might for example play the role of working hypotheses that help us to develop our position. Lastly, commitments can have credibility that is independent from the current RE process if (c) background theories support them (Baumberger and Brun 2017, 177), for example, in the context of justifying a theory of distributive justice, a commitment to help people who are much less well-off may be supported by a moral background theory which includes a requirement to help the poor.

As point (b) shows, a minimal degree of commitment is enough to qualify an input commitment to enter the process. That is, if we are committed to a consideration, this is enough to ascribe it some independent credibility and to let this commitment enter the process of adjustments. We do not need to identify additional (e.g., inferential) support for this commitment. In case of conflicts, these weak commitments will often be discarded most easily; however, it can also be that a number of weak commitments together outweigh a more "weighty" one (cf. Scheffler 1954, 182). Thus, admitting also very weak commitments—for example, based on a mere hunch—is one of the ways that opens the possibility for innovation in reflective equilibrium, while insisting that a consideration has to have a particular "pedigree" to be independently credible may lead to conservatism and entrench bias.

Consequently, even having a very high degree of independent credibility is not enough for the justification of a commitment (Elgin 1996, 102): it also has to be part of a position in which it and the rest of the relevant commitments are accounted for by a system that does justice to theoretical virtues—that is, in order to be justified, a commitment needs to be part of a position that is in reflective equilibrium.

This means that while the independent credibility of a commitment plays a role for justification, and has to be respected, it is not determinate. The question is not only how strongly we are committed to each of our individual commitments, but also how willing we are to commit to the whole set taken together—and this might sometimes speak against upholding one or more individual commitments. Compare the following quote by Scheffler (1954):

> The justification for accepting [a sentence] *A* at a given time may now be made not on the grounds of its own initial credibility, nor of some unspecified coherence, but on the basis of its coherence with the system which maximizes initial [i.e., in the terminology used here: independent; T.R.] credibility at that time, while, together with its sister sentences, *A* indirectly controls the choice of this standard system. (Scheffler 1954, 182)

Respecting input commitments and their independent credibility can indeed be a reason to reject specific input commitments, even those that have a high independent credibility: while the independent credibility of a commitment speaks in its favor, the independent credibility of other commitments can—in case of conflicts—speak against it (mediated through a system that accounts for some, but conflicts with other, commitments).

When adjusting commitments, we thus have to make plausible that the independent credibility of a commitment was not simply discarded, but in fact outweighed by other relevant considerations, i.e., to argue that independent credibility was respected adequately. Another way to put this is that adjusting the commitment in question should increase the credibility of the epistemic position *as a whole* more than the independent credibility of the commitment would contribute to it. Maximizing the credibility of the position as a whole is not the same as minimizing deviation from the independently credible starting positions (cf. Elgin 1996, 109).

Brun (2013, 241) gives the example of a supporter of gay marriage who appeals to a 'between consenting adults' principle and realizes that this conflicts with his commitment to the wrongness of polygamy. He can then, e.g., argue that the latter "was feeble anyway in comparison to the principle which covers many commitments he firmly holds". This would be an instance of referring to the relative weights of commitments and system, i.e., the other commitments outweigh the polygamy-commitment via the 'between consenting adults' principle. The epistemic position as a whole is more plausible without the commitment.

Alternatively, or additionally, he could argue that his views on polygamy were merely a cultural prejudice. This would mean that the independent credibility of the commitment cannot withstand scrutiny, and that he is actually *lacking* positive reasons to uphold the commitment. If his commitment turns out to be grounded only in cultural prejudices, he has no valid reasons that could explain why there should be a relevant difference between gay marriage and polygamy, if we are upholding a 'between consenting adults' principle. The independent credibility of the commitment is in this case not only outweighed, but *lost*.[4]

The criterion of independent credibility means that RE does not create justification "*ex nihilo*" (Baumberger and Brun 2017, 176)—instead of "pure" coherentism, it makes RE a combination of (weak) foundationalism and (weak) coherentism (cf. Hansson 2007). Justification by reflective equilibrium requires that the resulting position includes at least some commitments which also have some degree of credibility independently of the position. As Elgin (1996, 107) puts it, "For reflective equilibrium, independently motivated, initially tenable commitments must underwrite coherence."

One reason why a resulting commitment has independent credibility might be that it is an independently credible input commitment which survived the process of adjustments. But not only input commitments can have independent credibility: it is possible that commitments that are the result of adjusting an input

[4] See Elgin (1996, 109) for further examples of how independent credibility can vary, or be lost.

commitment have some credibility that is independent of their coherence with the current position. Consequently, respecting independent credibility does not only refer to the independent credibility of input commitments. Following Baumberger and Brun (2021, 7934), we can distinguish *independent* credibility, the credibility a commitment has independent of the coherence of the current position, from *initial* credibility, the credibility a commitment had at the initial stage.

Since not only input commitments can have independent credibility, and because independent credibility can vary and change over time, adequately taking independent credibility of commitments into account does not automatically ensure that we do not change the subject. The subject matter is constrained by our input commitments, and those are what have to be respected in order not to abandon the subject.

2.3.2 Not Abandoning the Subject

When we start a project of inquiry in order to search for a justified answer, we do this because we want to learn something about a particular problem, or particular subject matter. Thus, simply abandoning the subject and focusing on another problem will not be an adequate answer. Strawson (1963) has expressed the worry that we might be tempted to simply change the subject in the following way:

> [We] may be diverted from the wish to understand what we are doing, by encouragement to do something else; and [...] if the wish seems futile, the diversion may seem desirable; and then the complaint that the wish is not thereby satisfied will, not doubt, seem futile too. (Strawson 1963, 509)

Baumberger and Brun (2017) give two examples for how a process of reflective equilibrium could go astray and abandon the subject:

> If a course of reflection leads to a theory which merely underwrites *might is right*, it will not count as providing a *moral* theory since it would force us to give up too much of our most important moral commitments. Similarly, if a model turns out to describe only short-term conditions of meteorological variables such as temperature and precipitation in a given region, it will not count as a *climate* (in contrast to *weather*) model. (Baumberger and Brun 2017, 179)

However, even if we do not want to change the subject in this radical sense of *abandoning* it and ending up with a theory or systematization of another subject, it still has to be possible that the subject undergoes significant changes in the sense of *alterations* and *elaborations* (Elgin 1996, 130): we want to be able to get rid of prejudices and biases, we want to be able to include further relevant considerations, and we want to be able to explicate concepts so that they are more suitable for our epistemic objectives—even if this conceptual change can change the extension of our concepts (compare the famous example of explicating "fish" so that it excludes whales).

It seems plausible that making progress in understanding and theory-development can also consist in the insight that our original delimitation of the subject matter was not helpful or misled. Systematizing our commitments about a specific subject often requires considerable re-thinking and re-categorizing—potentially going so far that we will speak of "paradigm changes" (Baumberger and Brun 2021, 7934). Take the following example: using your moral commitments you decide that a course of action is impermissible because it conflicts with the principle that one ought not inflict gratuitous pain on people. You then notice that in that principle the term 'on people' does no work. The principle should be 'one ought not inflict gratuitous pain'. This immediately extends your moral theory to all animals capable of experiencing pain. Have you changed the subject of your moral theory? You have definitely massively extended its scope.[5] But arguably, you also made substantial progress by recognizing the moral value of non-human animals.

How, then, can the criterion of not abandoning the subject be spelled out in a way that still allows us to make progress? What we want to do is to improve upon our initial commitments (see, e.g., Elgin 2017, 66) with a specific (set of) goals, or questions, in mind. This means that in order to ensure that we stay on topic, it is necessary that we can trace how the resulting commitments were obtained from the input commitments. But this is not enough: we also have to be able to give reasons, or arguments, for why these adjustments consist in improvements of our original grasp of the subject matter. In short, we need to be able to give arguments that can explain why the input commitments seemed plausible, but still needed to be revised.

This "tie" back to the input commitments ensures that even if our position has substantially changed, we can still see how it is connected back to our initial understanding of the subject, and the initial problems we started out with. This requirement thus highlights the importance of the *process* of adjustments in RE: it is not just enough to describe the initial and the resulting position, but rather we need to be able to reasonably trace how the latter was constructed based on the former. We can read the following quotation from Strawson (1963) as an expression of this:

> [If] the clear mode of functioning of the constructed concepts is to cast light on problems and difficulties rooted in the unclear mode of functioning of the unconstructed concepts, then precisely the ways in which the constructed concepts are connected with and depart from the unconstructed concepts must be plainly shown. (Strawson 1963, 513)

This process can be a reconstruction and does not need to be a description of how we did actually proceed, but it needs to allow us to see how the resulting position can be reached starting from the input commitments. After all, we want to learn something about the commitments that we start out with, and to improve upon them—and learning why we should revise them is one way in which we might achieve this goal.

[5] I am grateful to Catherine Elgin for this example.

Which changes of input commitments are compatible with addressing the relevant subject matter is not a question that has a clear-cut answer, however. Instead, it has to be answered based on arguments referring to input commitments, the overall pragmatic-epistemic objective, and background theories (Baumberger and Brun 2021, 7933). In the example from Baumberger and Brun (2017) above, we did fail to formulate a moral theory, and we did construct a weather and not a climate model. These failures become apparent not only because we made too-drastic changes to the input commitments, but especially because we can argue that these projects missed their objective: our background expectations for a moral theory speak against the *might is right* principle, and background theories about the distinction between climate and weather allow us to argue that the constructed model fails to be a climate model.

Thus, it seems illusory to expect that there is one precise yet generally applicable criterion for whether or not the subject was changed. Generally, we should expect that there will still be some overlap between input and resulting commitments, that resulting commitments will accommodate some of the same evidence as input commitments did, or that the theory will still be applicable to the same cases and questions as the input commitments intended to answer. For example, when constructing a moral theory, we might accept that we were wrong in our judgment about whether or not a specific action is permissible, but we will still want to know what we should do in this situation. And we will expect that the answer stands in an appropriate relation to our input commitments about relevant factors for moral evaluation (cf. Brun 2020, 937–38).

The criterion of respecting input commitments in order to ensure that the subject is not abandoned is thus a different criterion than the one that commitments should have independent credibility. The independent credibility of a commitment—e.g., how it is supported by evidence or background theories—will of course typically play a role in the arguments for whether or not an adjustment, or series of adjustments, amounts to abandoning the subject. But it is not only the input commitments that can have independent credibility, and the not-abandoning-the-subject requirement is specific to input commitments (initial and emerging). Purely inferential commitments, that is, commitments the agent adopts solely because she inferred them from the system currently under construction, do not have independent credibility. Another way to say this is that the system cannot be a source of credibility that is independent of how well the commitment fits into the position under construction, as the system is itself a part of this position. What starts as a purely inferential commitment can gain independent credibility later, e.g., if it turns out that it can be supported by background theories. But even then, those commitments do not play a role for providing understanding of the original subject matter and consequently do not need to be respected like input commitments (Baumberger and Brun 2021, 7932; compare also the status of "merely derived beliefs" in Hansson 2006).

One last point, before we move on to theoretical virtues of the system: of course, not switching the subject is not an ultimate epistemological demand. An agent may very well abandon a subject and start to pursue other goals, if, e.g., she continues to fail to reach reflective equilibrium. However, if we do abandon the original subject, we should do so consciously. We start to speak about other problems if we are no longer able to show how our current inquiry ties back to the input commitments of the original problem. Consequently, our results will have no direct authority or justificatory power on the abandoned subject.

2.4 Theoretical Virtues

There are some epistemic goals that commitments and the system both should fulfill qua them being parts of the position, that is, the position should be internally and externally consistent. Additionally, as discussed in the section above, commitments should have independent credibility. In this section, I focus on the epistemic goals that are specific to the system, i.e., theoretical virtues.

As I have already mentioned in Sect. 2.1, the system should do justice to a number of theoretical virtues. Virtues such as accuracy, scope, simplicity, and fruitfulness will be relevant in most contexts (at least to some degree, and most likely in different interpretations). There are also more specific virtues, which will be relevant only in some contexts, e.g., visualizability for purposes of scientific modeling, or the ability to be action-guiding for moral principles.

These virtues will sometimes pull in different directions, for example, it might only be possible to gain simplicity at the cost of accuracy, or vice versa. Thus, they will also have to be weighed against each other—and against the criteria of respecting input commitments and the independent credibility of commitments. The specific pragmatic-epistemic objective that we pursue in the RE process in question informs which specific configuration of virtues is relevant (Baumberger and Brun 2021, 7928). For example, if we want to justify an approach to decision-making under uncertainty for public policy-making, other theoretical virtues (or at least another weighting of them) might be relevant than if our goal were a purely formal decision theory. In the second case, mathematical precision may be paramount, whereas in the first case, the ability to be action-guiding will be much more important.

Demanding that the system has theoretical virtues ensures its systematicity, since it blocks the strategy of writing down the list of commitments, calling it a system, and claiming to have reached a position in reflective equilibrium (Baumberger and Brun 2017, 177–78). Striving for theoretical virtues in order to be able to meet a certain pragmatic-epistemic objective is what drives the process of adjustments forward against the more conservative "pull" from the demands to respect input commitments and their independent credibility. The criterion that the system should have theoretical virtues thereby also contributes to the coherence of the position, for example, because it asks for a system that can categorize commitments, identify relevant features, and relate them to each other. Having theoretical virtues is thus a part of what makes a system *systematic*.

The idea that it is a pressure for systematization that drives us to move away from what we were initially committed to about the subject matter, that is, what drives us to make progress, is expressed nicely in the following quote by Scheffler (1954):

> [Justification] is the *systematic* rechanneling of initial commitments in such a way that each act is judged in terms of all others. We do not start from scratch, but always with initial commitments of some degree; but neither do we rest content with the latter. We modify and transform them into derived commitments of various sorts by *systematic pressure* which is channeled through principles of congruence. (Scheffler 1954, 188, italics added by T.R.)

There are two reasons why the RE criterion demands that the system should *do justice* to a configuration of theoretical virtues, and not *realize* them exactly. Firstly, as already mentioned in Sect. 2.1, theoretical virtues can pull in different directions, and sometimes trade-offs will be unavoidable. Demanding that a system reaches all the goals at the same time would thus be unrealistic.

Secondly, the exact characteristics of the configuration of epistemic goals, e.g., which virtues are relevant, how much weight they should have, and how they are exactly interpreted, is a result of the process and not a precondition for it (Baumberger and Brun, 2021, 7929; Elgin, 2017, 89). The configuration of theoretical virtues that we start out with is also a form of higher-order commitment to the standards that we employ in our inquiry. Like first-order commitments about the subject matter, they can change during the process. And like first-order commitments, theoretical virtues can be supported or undermined by background theories about specific theoretical virtues and their cognitive merits (Baumberger and Brun 2021, 7929; Elgin 1996, 105). Thus, like first-order commitments about the subject matter, our higher-order commitments in the form of theoretical virtues and other standards of inquiry are open to revision, but have to be respected during the process.

2.5 Background Elements and Social Dimensions

We have already seen that it is central for RE to reach a state of agreement between commitments and a systematic account of them. However, to be justified, it is not enough for a position of commitments and a system to be internally in agreement. The position should also be externally coherent, that is, with related theories and otherwise accepted and justified theories and conceptions (cf. Kuhn 1977, 323). Instead of settling for the best "fit" between commitments and system candidates, we should also refer to background theories in order to bring forward arguments for or against possible adjustments. The justification of a current position in the foreground thus takes place against a background that is relatively fixed and—at the given time—not called into question. This background necessarily will include epistemic achievements of others, as a single epistemic agent is not able to justify everything on their own.

2.5.1 *Background Theories, Background Information, and Background Assumptions*

Within the background, three rough distinctions can be made: firstly, there are accepted background theories which are vindicated in a way relatively independent of the foreground, that is, their justification should be relatively independent from the current project of inquiry. During the RE process, they will also guide the process of adjustments by providing additional support or constraints on the foreground system.

Secondly, often additional background information is needed to relate the system and the commitments to each other, e.g., in order to infer something from a moral principle on a specific action, factual information on the relevant situation in which the action takes place needs to be added:

> Moral beliefs often explain, entail or conflict with other beliefs only if some context of nonmoral assumptions is presupposed. For instance, hedonistic utilitarianism does not by itself conflict with, or entail, the view that it was wrong of USA to wage war against Iraq, but only in conjunction with claims about the consequences of USA's war. (Tersman 1993, 54)

Thirdly, in practice we will often have to make background assumptions, because we cannot investigate everything at the same time and will have to make some stipulations in order to get the process going. As long as these assumptions don't do the "real work" of systematizing the commitments, this seems relatively unproblematic, but it requires paying close attention to it while actually carrying out (or reconstructing) an RE process.

Sometimes, the background is seen as another element of the process of adjustments, which is to be adjusted mutually with the commitments and the system.[6] However, while I agree that nothing is, in principle, safe from revision in RE, I think the background should not be seen as one of the primary elements to be adjusted in a given project of justification. What is accepted as background also partly constrains the subject matter, and if we start to change the background too drastically, we will also change the subject. Another way to put it is to say that by *not* including what is in the background in our input commitments, we also express a sort of higher-order commitment about the subject matter in question.

---

[6] For example, Daniels (1979) names judgments, principles, and background theories as the elements that are to be adjusted.

Thus understood, the distinction between foreground and background is not an absolute distinction, but a matter of perspective: it depends on the specific pragmatic-epistemic project that we are pursuing. During an RE process, parts of the background can be called into question and become part of the foreground, but this will typically mean that we are changing our epistemic project: for example, because we might have noticed that our delimitation of the subject matter rested on problematic assumptions in the background, and that we just cannot make progress in our original project without addressing these problems.[7] Background theories that are justified by their own RE process are more likely to remain fixed through the process than background assumptions, since their status as being part of a justified position of their own gives us strong reasons not to revise them easily. In general, it might be a good strategy to try to hold the background constant for as long as possible. But this does not change that there are no absolute, once-and-for-all fixed points.

2.5.2 *Going Public*

When taking background elements into account, we typically have to rely on theories, distinctions, or claims that we cannot directly prove or justify ourselves in order to continue to pursue our current pragmatic-epistemic project. Having to justify all that for ourselves before coming back to the question we are concerned with would mean that we never arrive there.

As Elgin (1996, 116) argues, relying on others in RE amounts to more than calling on experts to "patch holes" in our position, e.g., in situations where we are lacking the expertise to estimate the independent credibility of a commitment. Ultimately, we have to rely on the "resources of the community" for even trivial facts:

> My tenable belief that trash is collected on Tuesdays needs no support from the experts. My own experience bears it out. Still, to have that belief, and to have experiences that count as evidence for or against it, requires knowing what trash is, what trash collection is, what Tuesdays are. Such facts are socially constituted and are imparted through socialization. Without the resources the community provides, I could neither formulate nor justify the belief in question. (Elgin 1996, 116)

Not all of these background resources have to be made explicit by an agent engaging in a pragmatic-epistemic project by employing RE, of course (nor would this be possible). But we have to keep in mind that we can justify our positions only with, and against, the broader background of the relevant epistemic communities, without being able to review everything for ourselves. This is not in itself something

[7] Arguably, such an outcome—realizing that the real problem lies somewhere else than we first thought—also constitutes an important epistemic achievement.

RE specific. However, through the background–foreground distinction, RE makes the point of relying on background resources especially salient.

Not only do we have to rely on the resources of others because we cannot justify everything on our own, we should also do so in order to "broaden the base" so as to control "for perspective and for eccentricity" (Elgin 1996, 117). The social-epistemological elements of RE do not only concern background theories and background information, but agents should also consider the commitments of others, in particular commitments of experts or people whose opinion on the subject matter is otherwise relevant (maybe because they are practitioners, e.g., the commitments of nurses and patients should be considered when developing principles of ethics for care in nursing homes).

RE is thus inherently holistic, which also explains why the justification it provides is preliminary: we should cast the net as wide as possible, and include as many relevant considerations as possible, but there is always the possibility that new information arises that unbalances our equilibrium.

2.6 Summing Up: Criteria of Reflective Equilibrium

To sum up, we may list the following conditions for reaching a position that is in a state of reflective equilibrium:

1. The resulting commitments and the system are in agreement;
2. The resulting commitments and the system are supported by background theories;
3. The system does justice to the relevant theoretical virtues;
4. The resulting commitments respect the input commitments adequately;
5. The resulting commitments have independent credibility; and
6. The resulting position is at least as plausible as all available alternatives.

Criterion (6) is important because the criteria (1)–(5) can be met to different degrees, and trade-offs among them are possible. Thus, through following different pathways and exploring different ways of systematization and adjustment, multiple, equally plausible positions might be achieved that are in reflective equilibrium to the same degree, i.e., overall meeting the criteria equally well. As Elgin (1996, 119) argues, this is not a problem for justification, but rather a question of practicability: as long as my position is at least as plausible as yours, none of us has a good reason to abandon our position and adopt the other. However, simply because there might be more than one maximally plausible position which is acceptable, it does not follow that a position which is less than maximally plausible—which is inferior to another relevant alternative—is acceptable (Elgin 1996, 143–145).

And even though it is possible that more than one justified position results, it is also possible that *none* is reached. Following Elgin (1996), I see reflective equilibrium as an imperfect procedural epistemology. There is no guarantee that an equilibrium is reached, and we may with reason believe that a system is in equilibrium when it is not (Elgin 1996, 14). The consequence of this, however, is not skepticism, but *fallibilism*: justification is inherently provisional, but no less reasonable (Elgin 1996, 15).

The Way Ahead In this chapter, I described the theoretical foundations of reflective equilibrium as an imperfect procedural epistemology that is weakly foundationalist and holistic. However, this theoretical conception needs to be specified in various ways in order to become applicable in the form of a method. In the next Chap. 3, I identify the various methodological decisions and challenges that need to be made in order to get from this theoretical conception to an applicable method. In particular, I will suggest ways in which the initial position can be described, and in which the process of adjustments and the evaluation of its results can be structured.

References

Baumberger C, Brun G (2017) Dimensions of objectual understanding. In: Grimm SR, Baumberger C, Ammon S (eds) Explaining understanding: new perspectives from epistemology and philosophy of science. Routledge, New York, pp 165–189

Baumberger C, Brun G (2021) Reflective equilibrium and understanding. Synthese 198(8):7923–7947. https://doi.org/10/ggkp4w

BonJour L (1985) The structure of empirical knowledge. Harvard University Press, Cambridge

Brun G (2013) Reflective equilibrium without intuitions? Ethical Theory Moral Pract 17(2):237–252. https://doi.org/10.1007/s10677-013-9432-5

Brun G (2016) Explication as a method of conceptual re-engineering. Erkenntnis 81(6):1211–1241. https://doi.org/10.1007/s10670-015-9791-5

Brun G (2020) Conceptual re-engineering: from explication to reflective equilibrium. Synthese 197(3):925–954. https://doi.org/10.1007/s11229-017-1596-4

Cath Y (2016) Reflective equilibrium. In: Cappelen H, Gendler T, Hawthorne J (eds) The Oxford Handbook of philosophical methodology. Oxford University Press, Oxford, pp 213–230

Daniels N (1979) Wide reflective equilibrium and theory acceptance in ethics. J Philos 76(5):256–282

Dutilh Novaes C (2020) Carnapian explication and ameliorative analysis: a systematic comparison. Synthese 197(3):1011–1034. https://doi.org/10/gmpp6s

Elgin CZ (1996) Considered Judgment. Princeton University Press, Princeton

Elgin CZ (2014) Non-foundationalist epistemology: holism, coherence, and tenability. In: Steup M, Turri J, Sosa E (eds) Contemporary debates in epistemology, 2nd edn. Contemporary Debates in Philosophy. Wiley Blackwell, Chichester, pp 244–255

Elgin CZ (2017) True enough. The MIT Press, Cambridge

Gertken J (2014) Prinzipien in Der Ethik. Mentis

Hansson SO (2006) Coherence in epistemology and belief revision. Philos Stud 128(1):93–108. https://doi.org/10.1007/s11098-005-4058-7

Hansson SO (2007) The false dichotomy between coherentism and foundationalism. J Philos 104(6):290–300

Knight C (2017) Reflective equilibrium. In: Blau A (ed) Methods in analytical political theory. Cambridge University Press, Cambridge, pp 46–64

Kuhn TS (1977) The essential tension: selected studies in scientific tradition and change. The University of Chicago Press, Chicago

Rawls J (1974) The independence of moral theory. Proceedings and Addresses of the American Philosophical Association 48:5–22. https://doi.org/10.2307/3129858

Rawls J (1999) A theory of justice. Revised Edition. Belknap Press, Cambridge

Scheffler I (1954) On justification and commitment. J Philos 51(6):180. https://doi.org/10.2307/2021776

Strawson PF (1963) Carnap's views on conceptual systems versus natural languages in Analytic Philosophy. In: Schilpp PA (ed) The Philosophy of Rudolf Carnap, Open Court: La Salle, pp 503–518

Tersman F (1993) Reflective equilibrium: an essay in moral epistemology. Lagerblads tryckeri AB, Karlshamn

Van Thiel GJ, Van Delden JJ (2009) The justificatory power of moral experience. J Med Ethics 35(4):234–237

Chapter 3
Specifying the Method of Reflective Equilibrium: A Methodological Framework

Chapter 2 described the theoretical foundations of reflective equilibrium (RE) as an epistemological theory of justification. How do we get from this theoretical conception to a *method*? That is, how are we to obtain a method of RE that can be applied by actual researchers and practitioners?

This chapter develops a methodological framework which shows how the RE criteria can be specified, and how guidelines for its application can be defined—that is, how an RE *method* can be specified. As we saw in the introduction, there are different ideas about the aims and purposes of RE, as well as different ideas about the agent(s) involved in its application. For the purpose of this book, the objective is, firstly, to spell out a method that can be applied by one epistemic agent and does not require group deliberation or similar group processes. (However, this agent is of course not isolated, and needs to consider the inputs and epistemic achievements of others.) Secondly, reflective equilibrium is treated as a method that aims at justifying principles, theories, or similar; and not, for example, for finding a justified consensus, or as a decision framework. These restrictions do not mean that it is not possible to spell out RE methods for other purposes. However, they give the case study a more stringent focus while laying the groundwork for further work on different ways to spell out RE.

With respect to these two restrictions, this chapter develops an RE *methodology*: it describes guidelines for how RE can be specified and applied as a method, in a way that allows imperfect epistemic agents to apply it without sacrificing methodological rigor. It is worth pointing out that most of the methodological issues that need to be resolved for this purpose are not inherent to reflective equilibrium as a theory of justification (Walden 2013, cf.). For example, the RE epistemology tells us that commitments should have independent credibility, and that to be justified, the resulting commitments and the resulting system should be in agreement. But it stays silent on what, exactly, are sources of independent credibility, and how we

T. Rechnitzer, *Applying Reflective Equilibrium*, Logic, Argumentation & Reasoning 27, https://doi.org/10.1007/978-3-031-04333-8_3

can measure whether system and commitments are "in agreement". Many questions that one needs to address to be able to obtain an applicable method of RE need to be answered by other research fields, and will be part of (epistemological) background theories. Like everything in RE, the answers to these methodological questions are not fixed once and for all. The goal of this chapter is thus not to spell out an RE method in the sense of defining all its elements and criteria as precisely as possible. Instead, it describes the tasks one has to complete in order to apply RE as a method, the methodological decisions one has to make in order to be able to address these tasks, and the challenges one faces when making these methodological decisions. Spelling out the method in detail for its specific application is part of Chap. 5, which describes the design and setup for the RE case study on precautionary principles.

To start, let us reconsider the criteria for an epistemic position to be in a state of reflective equilibrium:

1. The resulting commitments and the system are in agreement;
2. The resulting commitments and the system are supported by background theories;
3. The system does justice to the relevant theoretical virtues;
4. The resulting commitments respect the input commitments adequately;
5. The resulting commitments have independent credibility; and
6. The resulting position is at least as plausible as all available alternatives.

Notably, these criteria concern the *resulting* position. But how do we achieve such a position in RE? And can these criteria guide us in spelling out rules, or at least guidelines, for how to proceed when adjusting commitments and system?

As described in Chap. 2, the RE process starts from the initial commitments about a subject matter, and then systematically adjusts them with respect to a system (e.g., principles, a theory, or a model), which, in turn, is formulated and adjusted with respect to the commitments. The whole process is guided by theoretical virtues, like simplicity or scope, that put constraints on the system, while commitments have different weights that put constraints on which adjustments are admissible. Additionally, the position in the foreground should be supported by background theories.

Thus, if we want to apply reflective equilibrium, the first stage will consist in clarifying our initial epistemic position, which includes identifying commitments, relevant background theories, theoretical virtues, and candidates for the system (Sect. 3.1). Starting from this initial position, we will enter the stage of adjusting commitments and system alternately (Sect. 3.2), until the process comes to a preliminary end point. In the latter stage, we will then have to critically assess and evaluate the resulting position concerning whether and to which degree it meets the criteria for being in RE (Sect. 3.3).

3.1 The Starting Position of a Reflective Equilibrium Process

The main tasks of the first stage are (i) to clarify one's goal, that is, to formulate the pragmatic-epistemic objective that one wants to achieve by applying reflective equilibrium, and (ii) to clarify one's initial epistemic position, that is, to identify and select initial commitments, theoretical virtues, background theories and background information, and candidates for the system.

Being as clear as possible about one's pragmatic-epistemic objective will help to (preliminarily) identify and specify all the other elements that are needed for the RE process. Some possible examples for such objectives are: to find a principled answer to the question of in which cases one is allowed to lie, to formulate and defend a theory of justice, to construct a weather model that can be used for predicting the local weather in mountain valleys, to justify a theory of inductive logic, or to justify a precautionary principle that can guide climate policy decisions. Having a clear goal will help to constrain the subject matter, which in turn helps to identify relevant commitments, relevant background elements, and existing candidates for the system. It also helps to determine which theoretical virtues are (likely to be) especially important for the target system in order to meet the objective. As with everything in RE, though, the formulation of the pragmatic-epistemic objective is not a one-way street: as we learn more about the subject matter, we might be able to formulate our objective more precisely.

In the remainder of this section, I focus on the tasks, methodological decisions, and challenges that one needs to address when one wants to clarify and describe one's initial position in order to start an RE process. Of course, these tasks can influence each other and do not have to be done in a strict order.

3.1.1 Identifying Initial Commitments

To improve our epistemic position and to work towards a state of reflective equilibrium, we need to assess the degree to which our commitments and our system are in agreement. When assessing this relation of agreement and deciding about possible adjustments, the independent credibility of the commitments has to be considered, that is, the credibility that the commitments have independently of their status in the RE process.

When we want to apply RE, one important task is thus to identify relevant commitments and their independent credibility. This poses a number of challenges. Without claiming to be exhaustive, I address now what I consider to be the five most important challenges.

Making One's Own Commitments Explicit Firstly, the epistemic agent needs to find ways to make her own commitments about the subject matter explicit, at least as far as possible. This is not trivial, as we typically are not aware of everything that we are committed to, and even less the reasons why we hold these commitments.

Making one's commitments explicit will to some extent be a creative process, but can be guided by systematic considerations: the agent can familiarize herself with the relevant literature, consider relevant cases in order to see how she would decide, use brainstorming techniques, discuss with others, and so on.

The process of making one's commitments explicit will also typically be incomplete, and has to be continued throughout the process of adjustments. There is always the possibility that some new considerations emerge that were not considered before.[1]

Considering Commitments of Others Secondly, on her own, the epistemic agent may only come up with a limited selection of relevant considerations. To broaden the input, the agent should thus also consider the opinions of others, like other researchers, but also of practitioners and of affected parties—for example, in a biomedical context, the commitments of physicians, nurses, and patients (Van Thiel and Van Delden 2009). While the agent does not have to accept the commitments of others at face value, it is important to consider them in order not to overlook aspects of the subject matter that the agent herself might not have access to, or where the agent might be lacking the experience to form commitments. The agent thus has to identify relevant other parties whose commitments should be considered, and select appropriate methods to obtain them. What methods are appropriate depends on the project at hand, e.g., its objective, scope, and subject matter. Some examples are empirical studies, sending out surveys, studying relevant literature, interviews, and personal discussions.

Specifying Which Considerations Are Independently Credible Thirdly, it needs to be specified what kinds of commitments have independent credibility. There is no general, project-independent answer to the question what gives a commitment independent credibility, and how much weight should be assigned to a commitment because of it. Instead, these are parameters that have to be addressed for each particular application of RE. To do so, the epistemic agent will have to refer to epistemological background theories, e.g., theories of confirmation, perception, intuition, and testimony (Baumberger and Brun 2017, Fn. 16). Which sources of independent credibility are relevant and how much weight they should give a commitment also depends on the project at hand, e.g., whether it is an empirical or a normative research project. In addition to perception or intuition, which were already named in Chap. 2 as possible sources of independent credibility, Van Thiel and Van Delden (2009, 236–37) name durability, transcendence, and experienced perception.

[1] In some situations, there might even be barriers that keep an agent from making her commitments explicit, e.g., if the agent is lacking relevant concepts to describe her experience, like in cases of hermeneutic injustice (Fricker 2007).

Durability means that we should have more confidence in commitments to judgments that are confirmed in a history of cases.[2]

Transcendence refers to the extent to which commitments are appreciated and affirmed by a community.

Experienced perception means that we should give more weight to commitments of people who have relevant experience, e.g., the commitments of medical practitioners in contexts of medical ethics.

Two things are important to stress here: firstly, as explained in Chap. 2, for a commitment to count as independently credible, we do not have to be able to give positive reasons for a commitment above and beyond that *we are committed to it*, i.e., are endorsing it and acting according with it in our actual practice. Often, we will have additional reasons to support it, but this does not have to be the case. Ultimately, whether or not such a commitment can be justified will depend on whether or not it can be shown to be part of a resulting position in reflective equilibrium. Secondly, no matter how the independent credibility of a commitment is identified, it is only a factor that gives additional weight to commitments—in no way does it make commitments safe from revision.

How to Individuate and Count Commitments? Fourthly, the question of how commitments can be individuated and counted poses a challenge for listing commitments. Tersman (1993, 44) argues that it is highly dubious whether there is a useful way of counting and individuating the beliefs of a person—it is far from clear whether the commitments of a person can be represented by a determinate and finite set of propositions. Tersman concludes that this suggests that the question to what extent commitments and system are in agreement must be assessed intuitively.

Let us look at an example to see why it poses a challenge for the assessment of agreement, and for deciding how to adjust one's position. Let us assume that I have a broad range of commitments of the sort "When coming home late from being out with friends, it is wrong to lie to your partner and say that you had to work", "You must not lie to your mother and tell her that you are sick when in reality you just do not want to go to her birthday party", "When you borrowed a book from a friend and lost it, you must not tell him that it was stolen", etc. Based on these commitments, I adopt the principle "It is always wrong to lie", which is in agreement with 25 commitments of this sort. However, I also have one single commitment expressed in the sentence "When you know that person *A* wants to murder *B*, then you should tell *A* that you do not know where *B* is, even when you are sure about the whereabouts of *B*." This latter commitment conflicts with the principle that it is always wrong to lie. How should we resolve the conflict?

[2] Cf. Elgin (1996, 102): "If projects grounded in a particular judgment often go awry, reservations develop and the courage of that conviction wanes. So confidence in a given judgment indicates that we have not yet found it an impediment to action. And that its acceptance has not obviously thwarted (and may even have advanced) our efforts is a reason to credit a judgment."

Note first that, in reflective equilibrium, this problem is independent of the method of counterexamples. We have a supported claim 'It is always wrong to lie' and an independently plausible case that is in tension with it. That just sets a problem. The mere fact that there is a counterexample to the principle doesn't determine whether we should call for a revision of the principle, or whether we should adjust or even reject the conflicting commitment.[3]

So how can we determine how to resolve the conflict? As I already mentioned, the individuation of commitments seems relevant for such problems. On the one hand, we could argue that the conflicting commitment is outweighed by the sheer number of other commitments that agree with the principle and thereby lend support to the principle over the conflicting commitment. On the other hand, we could also try to come up with even more commitments that are similar to the conflicting commitment, thereby increasing the number of commitments with which the system conflicts. Should we then say that the system is outweighed by the huge number of conflicting commitments?

One might think that a solution could be to somehow count commitments based on the relevant features they refer to. In the lying case, we could try to distinguish between classes of commitments where lying does negatively affect others, and cases where it would protect them. However, this raises the question of whether it is even possible to classify commitments *before* we have a system—after all, it is the purpose of the system to identify relevant features of commitments and to systematize them. Claiming that there is a relevant difference between these situations of lying is already a proposal for systematizing the commitments.

Possible ways to deal with this challenge could be to work with a fixed set of sentences that represent commitments, and to accept that the justification reached will be relative to this fixed set (as is the case, e.g., in the formal model of Beisbart et al. 2021). Another way is to start with a relevant selection of explicit input commitments, but to be aware that they are only a part of the whole set of commitments, that there are further implicit input commitments which can emerge during the process of adjustments, and to systematically search for them when adjusting system and commitments.

When assessing agreement between commitments and different candidate systems, it will be important to compare candidate systems with respect to the same set of commitments, and to search systematically for potentially conflicting or supporting commitments that need to be made explicit. This does not solve the problem of how to individuate and count commitments, but at least helps to ensure that not one candidate is rejected on basis of counterexamples that are not considered for another candidate. It also means that a resulting position can be defended as being in a state of RE only with respect to the commitments that were explicitly considered, and that it is always possible that there will be further emerging commitments that might or might not unhinge it.

[3] I am grateful to Catherine Elgin for helping me to make this clearer.

Making a Selection of Initial Input Commitments This means that, fifthly, a selection of those initial input commitments that should explicitly enter the process of adjustments needs to be made. One objection against reflective equilibrium is that too much—actually all that is relevant—depends on the selection of the input: from implausible or repugnant commitments can result implausible or repugnant principles (Kelly and McGrath 2010).

As an answer to this objection, some RE proponents propose to filter the commitments in order to only admit credible ones. This is in line with the demand of Rawls (1999, 42) that only those judgments in which we have high confidence, and which are made under "conditions favorable to the exercise of the sense of justice, and therefore in circumstances where the more common excuses and explanations for making a mistake do not obtain." Not just any judgment should be included, but only our *considered judgments*. For example, Beauchamp and Childress (2013, 405–09) argue that we should start with those moral beliefs that are part of the "common morality" and made by "moral judges" that have relevant epistemic virtues, such as being impartial, or having sympathy and compassion for the welfare of others.

To arrive at a justified output, the argument goes, it is important to start from credible inputs (cf. Van Thiel and Van Delden 2009). And indeed, one of the criteria of the RE conception described in Chap. 2 is that commitments should have independent credibility. Does this mean that we should "filter" our commitments in order to only include those that have a high independent credibility?

The RE conception from Chap. 2 does not presuppose a strong filtering condition for input commitments: commitments should have independent credibility, but to be included in the process of adjustments, this credibility can be very minimal—including the limiting case where nothing speaks against this commitment. Commitments are *considered* during the process, and considering and "vetting" or "filtering" them in piecemeal fashion prior to entering the process of reflective equilibrium, is, as Knight (2017, 49) argues, counterproductive: "we have no way of knowing whether these isolated speculations will be consistent with the most plausible overall position."

Van Thiel and Van Delden (2009, 235) also argue against a strong filtering condition in the sense of only considering highly credible commitments, because this might (a) lead to excluding intuitions from minority groups of agents, and (b) narrow down the available input too much—e.g., selecting commitments based on whether they belong to the "common morality", which is a set of norms all morally serious persons share, may result in a very small set of input commitments that does not provide a sufficient basis for moral theorizing. Furthermore, (c) "limiting the set of moral intuitions in this way complicates the task of integrating the relevant moral experience of others. For example, from agents who may have moral intuitions that are not shared by all morally serious persons, because their intuitions stem from moral experience that is uncommon" (Van Thiel and Van Delden 2009, 235). Only considering input commitments that we take to be highly credible might also (d) lead to conservatism, since we will be wary of altering them.

All this speaks in favor of also accepting commitments with a very low independent credibility, and to explore how the various commitments and their relative weights work together. Since (candidate) systems should account for all current commitments, making the set of commitments as broad as possible will enrich the process by adding more (potentially) relevant factors that might or might not be a suitable basis for systematization. Making sure that their weights are considered adequately means that commitments with a very low credibility will not easily lead to the elimination of more credible commitments (after all, that independent credibility of commitments has to be respected is a criterion of RE). At the same time, the low independent credibility of a commitment can be overridden by its being incorporated into a credible position—this might even be true for commitments that are initially "incredible" (cf. Elgin 1996, 119).

But if commitments are not selected based on their degree of independent credibility, how can we identify those commitments that we should explicitly consider in the process of adjustments? Since it is not possible for human epistemic agents to consider all and every commitment that they hold, a selection has to be made, even if one rejects a strict filtering condition (Van der Burg and Van Willigenburg 1998, 4). Firstly, it is important that this selection is made in a way that other researchers are able to understand why certain commitments were selected, even if they perhaps would have selected others (de Vries and van Leeuwen 2010, 498). Secondly, this highlights the role of "emerging" input commitments for practical applications: the epistemic agent should not only start with a specific selection of commitments, but should strive to broaden the set of relevant commitments throughout the whole process (Van Thiel and Van Delden 2009, 235). This can happen through systematically searching for counterexamples (or further supporting examples), through engaging with the literature, through considering thought experiments, through presenting and discussing one's commitments with others, etc.

It makes sense to start with a selection of what we take to be commitments to "core cases", "central problems", and "paradigm examples" of the subject matter. We also have general commitments about what does or does not belong to the subject matter, and should name some examples that we take to be especially relevant. They can of course also be adjusted, and, depending on the pragmatic-epistemic objective, sometimes it might be better to deviate more from the initial subject matter to further our epistemic goals, and sometimes less. It just needs to be "traceable back", so that it is plausible that we solved the problem we originally set out with. This might include recognizing that the problem is better described in another way than we originally thought—as long as we ensure that we did not abandon the problem altogether without such reasoning, i.e., did not simply leave it because it was too hard, or uninteresting, and went doing something else instead (which can be also legitimate, but in such a case we should not claim to still be talking about the initial problem).

Tasks to Identify Initial Commitments

- Identify relevant literature and other information sources on the subject matter and familiarize yourself with it;
- Employ strategies to elicit your own commitments concerning the subject matter;
- Identify relevant parties whose commitments should be considered and select appropriate methods to obtain them;
- Identify sources of independent credibility (based on epistemological theories and on what is relevant for the subject matter at hand);
- Make a selection of explicit initial input commitments that you deem representative, and explain your reasons for making this selection in a way that is comprehensible to other researchers.

3.1.2 Selection of Theoretical Virtues

To be part of a position that is in reflective equilibrium, the target system needs to have theoretical virtues—we want a *systematic* account of the relevant commitments, and this means that the target system—be it a (set of) principles, a model, or a theory—needs to have certain virtues like being simple, fruitful, or having unifying power. Which theoretical virtues are relevant, and how they should be weighted, depends on the overall pragmatic-epistemic objective that is pursued. For example, if you want to come up with a general moral theory—e.g., a theory of justice—the virtue of "broad scope" might be more important than if you are addressing a specific issue of applied ethics in a specific context, e.g., ethical questions concerning the patenting of genetic resources and ownership of digitized sequence information. Making a preliminary selection of theoretical virtues will thus help to further clarify what kind of project is pursued. Being as clear as possible about how one understands these virtues will help when comparing candidate systems in order to argue in favor or against them.

The virtues and their interpretation are not fixed once and for all, but can be seen as second-order commitments (Elgin 2014) that are also open to revision if, e.g., it turns out that prioritizing the virtue of broad scope is no longer contributing to our pragmatic-epistemic objective.

Nonetheless, to describe the initial setup of the RE process, it is necessary to make a selection, even if it is preliminary.

Making a Preliminary Selection of Theoretical Virtues

- Which theoretical virtues are likely to contribute to the pragmatic-epistemic objective? That is, which theoretical virtues are likely to be relevant for the target system given the pragmatic-epistemic objective?
- Which of those virtues are likely to be more important for the target system, and why?

3.1.3 Description of the Background

To be justified according to reflective equilibrium, the epistemic position of an epistemic agent should not only be internally coherent in the sense that commitments and system are in agreement. To be fully justified, this position needs to be in reflective equilibrium with respect to the best available background theories and to all available background information that is relevant for the subject matter (Baumberger and Brun 2017, 174–75).

For an epistemic agent who wants to apply RE, this poses the challenge of getting an overview of the background which is comprehensive enough without requiring processing so much information that the method becomes unworkable. As with the selection of initial input commitments, it might make sense to work with a relevant selection, while being aware of, and familiar with, a broader picture. Background theories and background information is also one aspect of applying RE where its social aspects become especially salient: even if a single epistemic agent is applying RE, they will have to rely on the epistemic achievements of others in order to identify justified background theories that they can use.

Which theories will be relevant to, e.g., constructing arguments in favor of or against commitments or candidate systems cannot be fully determined before starting the process. But thinking about it, considering relevant literature, and consulting others, will again help to get a clearer idea of one's subject matter: what are the theories that I presuppose and that I do *not* aim to justify as part of the position in the foreground? Which theories have sufficient justification to be used as external support for my position? For example, in the case of thinking about ownership of digitized sequence information, relevant background theories might concern ownership in other contexts.

That these background theories are, for the purpose of the RE process, *presupposed*, does not mean that they are safe from revision: just that they are currently not in the focus of what is being justified. It is possible that the process makes clear that one of these background theories has problematic implications and should be revised, but as soon as we do this, we are changing our epistemic project.

Identifying Relevant Elements in the Background

- What are relevant and plausible background theories?
- What information, e.g., factual information about the subject matter, is relevant?
- Are there any assumptions that need to be made?

The answers to these questions will also depend on the selection of initial commitments and on possible candidates for the system, e.g., what background theories are needed in order to correctly interpret them.

3.1.4 Selection of Candidate Systems

To get the process of adjustments going, we will need candidates for the systematization of the commitments in form of principles or a theory. Through familiarizing oneself with the subject matter and debates concerning it, one will most likely be able to draw up a list of the main contending candidates for the system. Further candidates should be added, if one can think of any that seem plausible (Knight 2017, 57). If there are no available candidates, one will have to formulate them in order to get the process going. This is largely a creative process which cannot be guided by explicit rules, but it will be helpful to think about the theoretical virtues that one is aiming for, and to examine the initial commitments to see whether some of them might be suitable to be reformulated as parts of the system. The selected candidates should be real alternatives, not just straw men, while at the same time the selection has to be kept at a manageable number. As part of this, it is particularly important to also consider candidates that one does not agree with, but that are, e.g., widely endorsed.

Selecting Candidates for the System

- What are existing plausible candidates from, e.g., the literature?
- Are there commitments in the set of initial input commitments that could be suitable as a (part of) the system?
- Can you come up with further plausible candidates for the system?

We did now address challenges for describing the initial position, and for identifying the elements that enter the process of adjustments. But, if we want to make progress from this initial position, we need to know how to bring virtues, weights, etc. to bear on the process of adjustments. We look at this in the next section.

3.2 The Process of Adjustments

An epistemic position that is in a state of reflective equilibrium is reached through a process of mutually adjusting commitments and systematic elements (see Chap. 2). This process of adjustments is often described in terms of moving "back and forth" between commitments and system—neither side takes priority nor is safe from being adjusted (cf. Goodman 1983, 64; Rawls 1999, 18; Cath 2016, 214).

The goal of this stage is thus to adjust commitments and system with respect to each other in order to maximize the RE criteria. The question is whether we can spell this out in the form of rules, or at least guidelines, that can be applied methodically. Can we structure the process of adjustments in a way that allows users of RE to conduct it systematically, and that will help them to make progress?

3.2.1 Steps of the Equilibration Process

While in principle nothing is safe from revision in RE, we cannot justify everything at once. When spelling out the methodical steps of the RE process, I thus work with the background/foreground distinction in the sense that the justificatory project takes place in the foreground, and adjustments should primarily made to the commitments and the system. The goal is neither to justify the elements in the background, nor the theoretical virtues, nor the sources of independent credibility (unless, of course, this is our pragmatic-epistemic objective, but in which case it constitutes the project in the foreground). The idea is that there is something that is adjusted—there is a specific objective with respect to which we want to make progress—and other things that are preliminarily taken to be fixed. (With a special emphasis on *preliminarily* fixed, for the purpose of a given justificatory project—they can always come into focus in other projects.)

Thus, I propose to adjust commitments and system in two alternating steps, in which either the commitments are adjusted with respect to the current system, or the system is adjusted with respect to the current set of commitments. As we want to make progress not only with respect to agreement between commitments and system, but also with respect to the other RE criteria, they should come to bear on the decisions about which adjustments should be made. I thus propose two general kinds of steps.

Two Kinds of Steps of the RE Process

Adjusting Commitments Keeping the system constant, find the set of commitments that maximizes the combination of (i) agreement with the current system, and (ii) independent credibility, and (iii) respect for input commitments, and (iv) support from background theories.

Adjusting the System Keeping the current set of commitments constant, find a system that maximizes the combination of (i) agreement with the current system, and (ii) theoretical virtues, and (iii) support from background theories.

For reasons of cognitive manageability, it is also possible to hold parts of the position constant and only adjust a subset of the commitments with a part of the system. This also leaves room for, e.g., explicating concepts as part of an RE process (Brun 2020). In fact, it is to be expected that in practical applications, steps of the RE process will often consist of explicating concepts or systematizing subsets of the current commitments: in most cases we do not have (m)any pre-developed, fully fleshed-out candidate systems that we can compare. What Brun (2020, 934) says about explication also has consequences for RE:

> [In] many projects of explication the target theory is not readily available, and explicating concepts must therefore go hand in hand with developing a target theory[.]

Like explicators, RE agents typically cannot draw on a pre-developed system, and this is because developing such a system is one of the objectives of the RE process. But the building blocks of such a system, e.g., explicated concepts, are typically not readily available. An RE process might proceed in steps of partial systematizations which in turn impose further constraints on the continuation of the process. Brun (2020, 938–39) illustrates this with an example from Goodman (1951):

> [We] can imagine starting with an explication of *point* in terms of intersecting lines; this will have consequences for a subsequent explication of *to the left of*, the extension of which will now need to include certain pairs of pairs of intersecting lines; and this explication will in turn restrict our choice of explicata for *to the right of* to converses of the relation *to the left of*; and so on for further explications.

Thus, the two alternating steps of the RE process can be carried out both with the current system/set of commitments as a whole, or with parts/subsets of them. The important point is that if a "sub-process", e.g., an explication, is conducted, this always has to happen with respect to the overall pragmatic-epistemic objective for the position as a whole.

A challenge that becomes immediately clear is that the different criteria that should be maximized will often pull in different directions in both kinds of steps. What if one way to adjust commitments leads to a set that has a higher degree of agreement with the current system than another set, but the latter has a higher degree of independent credibility? How should such trade-offs be decided? And how in the first place can we assess whether or not a set of commitments has a higher degree of agreement than another?

Thus, the challenges for this stage are to specify the criteria of the two steps in a way that allows us to assess the extent to which they are met, and to decide how situations should be resolved in which trade-offs between criteria need to be made.

I first focus on the specific challenges for defining and assessing the RE criteria as part of specifying the method, before focusing in particular on the question of how trade-offs can be handled (in Sect. 3.2.3).

3.2.2 *Defining and Assessing the RE Criteria*

To specify the two alternating steps proposed above, we need to find ways to measure and compare the following:

- Agreement between (sets of) commitments and candidate systems;
- Support from Background Theories;
- Theoretical virtues of candidate systems;
- Respect of current commitments for input commitments.

On this basis, we can then address the issue of how to handle and decide trade-offs between agreement, theoretical virtues, and weights of commitments.

In the following, I describe what challenges need to be addressed in order to specify these measures. How I specify them for the purpose of my case study on precautionary principles is part of Chap. 5.

3.2.2.1 Measuring Agreement

The central relation that has to be brought into equilibrium is the one of agreement between commitments and a system. Agreement can be spelled out in different ways, but is typically understood to require more than mere consistency. One way is to spell it out in the form of *account*, i.e., that it has to be possible to obtain the commitments via inference from the system (Beisbart et al. 2021). These inferences can be specified as deductive or also as non-deductive inferences, and they will typically require that the system is applied to relevant background information. For a full agreement between system and commitments, we might require that the system can account for all of the commitments, and that we are committed to everything that follows from the system.

When measuring the account between the current commitments and the system, we face at least two distinct challenges:

Firstly, not all input commitments are likely to be explicit in the current commitments. This means that when comparing different candidate systems with respect to their ability to account for commitments, their success will be contingent on which commitments are explicitly considered.

Secondly, we will typically not have an overview of everything that does follow from a candidate system, that is, we will not have an overview about its logical closure.

Both challenges mean that it is often difficult, if not impossible, to get an accurate measurement of agreement (unless we work with a restricted, fixed set of inputs, like the dialectical structures used in the formal model of Beisbart et al. 2021). Thus, it is possible that a candidate system S_a is accepted over another S_b on the basis that it can better account for current commitments, while it later might turn out that there were actually further—implicit—input commitments that S_b can account for while S_a actually conflicts with most of them. Additionally, even if we decide to assess account only with respect to the commitments that are currently made explicit, it will still not be possible to compare those commitments with everything that follows from the system if we cannot overview all of it.

Possible ways to deal with these challenges are to keep these limitations in mind when applying the two steps, and to use heuristics to compensate to some degree for these shortcomings: thus, we should always make explicit with respect to which set of commitments agreement is measured, and we must systematically search for commitments that are not made explicit yet but might cause problems, as well as for inferences from the system that might lead to conflicts.

Defining a Measure for Agreement

- Decide how explicitly and exactly agreement should be measured (this also depends on what is feasible within the project at hand);
- Specify different degrees of agreement (e.g., account, consistent non-account, conflict), and decide how they should be weighed.

3.2.2.2 Measuring Support from the Background

To be in reflective equilibrium, the position should not only be internally coherent—agreement between system and commitments—but should also be reasonable given all of the best available background theories and background information. Given that one is aware of the background, it seems likely that support from background theories could be measured analogously to agreement. However, this will quickly become very complex and unworkable for a single epistemic agent, who would have to be familiar with all relevant background theories and their implications. In most applications, it will thus be more sensible to use references to the background selectively, e.g., as tie-breakers in cases of trade-offs, and to assess whether already well-developed positions can reasonably be seen to be in a state of reflective equilibrium (see Sect. 3.3). Assessing the degree to which a position is supported by background theories will typically have a strong social component, requiring that the RE applicant relies on the epistemic achievements of others who, e.g., justified said theories.

Defining a Measure for Support from the Background

- Decide how explicitly background elements should be included, and at what point(s) support from them should play a role.

3.2.2.3 Measuring Theoretical Virtues

The system does not only have to be adjusted with respect to the (current) commitments, but also with respect to theoretical virtues: in order to provide a *systematic* account of the commitments, i.e., an account that helps us to meet the pragmatic-epistemic objective that we pursue in conducting the RE process, the resulting system should have theoretical virtues like scope or simplicity. The (preliminary) selection of a set of virtues was already addressed in Sect. 3.1. But if we want to be able to compare different candidate systems with respect to their theoretical virtues as part of the process of adjustments, we also need a way to measure and comparatively assess the virtues.

This poses a challenge, as it seems implausible that all relevant theoretical virtues will always be operationalizable on a ratio scale or even an interval scale (Stegenga 2015, 269). Okasha (2011, 102) illustrates this with the example of the virtue of "fruitfulness":

> Conceivably, one could order a set of theories by how fruitful they are, but it is hard to believe that *differences* in fruitfulness can be compared; a statement such as 'the difference in fruitfulness between T1 and T2 exceeds the difference between T2 and T3' hardly seems meaningful. If this is right, then the real-valued 'utility' function that represents the fruitfulness preference order is merely ordinal—any increasing transformation can be applied to it without loss of information.

Different measurement scales may be appropriate for different criteria, and may also depend on the scope of the project that we are pursuing (cf. Okasha 2011, 103). Thus, for other virtues, it might be possible to measure them on ratio or interval scales. But as long as even one virtue remains that is measured on an ordinal scale, no complete ordering will be possible that meets reasonable conditions of theory choice (Okasha 2011, 93), because it is not possible to establish unambiguous trade-offs (Kemp and Grace 2010, 401).

When defining measures for the theoretical virtues, it thus makes sense to define them as precisely as possible, but to keep the pragmatic-epistemic objective in mind. Often, it might be better to use an ordinal scale instead of trying to force everything onto an interval or a ratio scale. However, virtues should at least be comparable on an ordinal scale, as theoretical virtues that do not allow for *any* comparison of candidate systems are not feasible for the RE process (and neither for any other purpose that aims at making a selection).

Defining Measures for the Selected Theoretical Virtues

- For each virtue, define a measurement scale that is at least ordinal.

3.2.2.4 Measuring Respect for Input Commitments and Independent Credibility

In Chap. 2, I argued that two senses of "respecting input commitments" can be distinguished: (1) as meaning that we have to respect their independent credibility, a criterion that also extends to adjusted or newly inferred commitments insofar as they have independent credibility, and (2) in the sense that, referring to the pragmatic-epistemic objective and the background, we have to be able to argue that the resulting commitments do not constitute a radical change of subject when compared with the input commitments.

For measuring, aggregating, and comparing the independent credibility of commitments, similar challenges might arise as with measuring the theoretical virtues: it might not always be possible to measure different sources of independent credibility—e.g., durability, experienced perception, or strength of intuition—in a way that allows us to obtain a complete ordering of (sets of) commitments. As with measuring virtues, it will make sense to be as precise as possible while ensuring that the measurement is still meaningful.

Concerning the question of whether the subject was changed, one way to interpret this criterion might be as demanding that as few as possible of the input commitments are adjusted. This could give us a negative measurement of "respect", where penalties could be assigned for differences between the input commitments and the current, adjusted set of commitments. The weights of commitments (i.e., their independent credibility) could be integrated into this measurement by, e.g., increasing the penalty for eliminating a commitment with a higher weight (cf. Beisbart et al. 2021).

However, we could also argue that while the input commitments constrain the subject matter in some ways, they do not define it completely. For example, I might be committed to the proposition "We should not research and develop climate engineering technologies to alleviate adverse effects of anthropogenic climate change". If I adjust this commitment in the sense of reversing so that I am now committed to the proposition that we *should* research and develop these technologies, did I really change the subject? This suggests that assessing whether the subject matter was changed in a (too) radical way is more a pragmatic criterion. Assessing it at every step might not be particularly useful—if we assume that the commitments constrain the subject matter, and respect their independent credibility when adjusting them, then this might be enough of a guide for the process. Whether or not the subject matter was changed too radically, and whether or not we were successful in achieving our pragmatic-epistemic objective, might better be assessed retroactively, once we have reached a (preliminary) end point of the equilibration process.

After addressing the challenges we face when defining measurements for the individual criteria, let us next focus on the question of how to handle the trade-offs that can result.

3.2.3 *Handling Trade-offs and Path-Dependency*

When adjusting the position in the foreground—commitments and system—we want to make adjustments in such a way that we make progress with respect to maximizing the criteria. The goal is to find a resulting position that is in a state of reflective equilibrium. This includes the requirement that it is at least as plausible as relevant alternatives, i.e., we need to be able to comparatively assess positions. I defined two steps of the RE process, in which either commitments or the system are adjusted with respect to the RE criteria:

Adjusting Commitments Keeping the system constant, find the set of commitments that maximizes the combination of (i) agreement with the current system, (ii) independent credibility, (iii) respect for input commitments, and (iv) support from background theories.

Adjusting the System Keeping the current set of commitments constant, find a system that maximizes the combination of (i) agreement with the current system, (ii) theoretical virtues, and (iii) support from background theories.

The discussion of how the individual criteria can be measured has revealed (a) that, often, criteria might only be measurable on ordinal scales, in which case forming a weighted sum (i.e., an unequivocal measure for which set of commitments, or which candidate system, maximizes the criteria) is not possible, and (b) that there can be trade-offs between the criteria which cannot be resolved unambiguously, i.e., where it is not clear which adjustments lead to an overall better position.

We might still think that there is one fundamental constraint for selecting a candidate position (or, in the specific steps of the RE process, a candidate system or a candidate set of commitments): whenever there is a candidate that is **pareto optimal**, i.e., that is at least as good as all other alternatives with respect to all criteria, and better in at least one, then this candidate should be chosen. When a position P_1 ranks as well as another Position P_2 with respect to, e.g., theoretical virtues and respecting input commitments, and ranks better with respect to account, then choosing P_2 anyway simply does not seem defensible. However, during the process, we need to be aware of the danger of local maxima: maybe sometimes we need to accept temporary epistemic setbacks in order to make overall progress.

In any case, it is unlikely that there will always be such a dominant option. What if P_1 and P_2 were equally good with respect to account, but the system of P_1 is more theoretically virtuous, whereas the set of current commitments of P_2 respects input commitments to a higher degree? Or when comparing candidate systems: what if a candidate S_1 is more fruitful than another S_2, but the latter has a broader scope (assuming that these virtues can only be measured on ordinal scales)?

Additionally, the weighing or even the selection of specific criteria might change during the process. For example, in the earlier stage of an RE process, it might make sense to trade off agreement for an increase in scope, because this will force us to critically examine more commitments and thereby works against conservatism. One reason why a system cannot account for a commitment might be that the latter is not in its range of applicability. We might then, in the early stages of an RE process, prefer a candidate system that has a broader scope, but conflicts with more commitments (i.e., fares less well with respect to the criterion of agreement), over one that has a narrower scope, but conflicts with fewer commitments (i.e., that fares better with respect to agreement). If, however, during several steps of the RE process, we simply cannot get rid of conflicts with central commitments, then we might want to reconsider, and trade off scope in favor of agreement.

Similarly, we might be more reluctant at the beginning to adjust commitments before having explored several candidate systems. Even if the current system, e.g., was the best available candidate in the step before, we cannot be sure whether there is not a much better candidate to be had, which we just did not come up with yet. Thus, adjusting with respect to the "weak" candidate would lead us astray. This does not have to be a negative thing—exploring different ways in which the position can be adjusted, and making each such pathway as strong as possible constitutes an important epistemic achievement, and also contributes to the justification of the position that ultimately results, because we can then truly show that it is at least as good as all alternatives. But, e.g., cognitive limitations and time-constraints will typically make it unfeasible to explore every possible pathway. At least when coming up with a first systematization, we will want to focus our energy on pathways that look somewhat *promising* (even if there is never a guarantee that we will not find still another, better candidate). This means that in the early stages of the process, it might make sense to put up with the conflicts in the position, and only to start adjusting commitments once a candidate system has in some sense "proved its worth", e.g., by being the strongest candidate in several rounds—or if actually all available candidates would conflict with the commitment. And of course, whenever we adjust a commitment that has independent credibility, we have to be able to give reasons why this adjustment is warranted—as long as we are not able to do this, we can at best tentatively explore whether adjusting this commitment in this way would, overall, lead to a position that is more defensible than a position in which the commitment was not adjusted in this way.

Thus, while in general we want to maximize the criteria, there does not seem to be a one-size-fits-all solution for trade-offs, and they rather have to be decided on a case-by-case basis with respect to the overall objective. The process of adjustments is neither infallible in the sense of guaranteeing that it leads to a state of reflective equilibrium, nor is it mechanical (Bonevac 2004, 386).

Specifying the Two Steps of the RE Process

- Define the RE criteria as precisely as possible while keeping them informative enough given the project at hand (i.e., the subject matter and pragmatic-epistemic objective in question, as well as the available resources);
- Give a preliminary weighting of the different criteria, noting whether any of them are more important with respect to the pragmatic-epistemic objective;
- Concretize the two alternating steps of the process by inserting the criteria so-defined.

3.3 Preliminary Conclusion of the Process and Evaluation of the Resulting Position

The process of adjustments comes to an endpoint when the position stabilizes, that is, when neither of the two steps improves the position anymore. But this does not yet guarantee that a full reflective equilibrium was reached. We thus have to assess to what degree the criteria are met in order to appraise the resulting position. This will include asking the following questions:

- Are the resulting commitments and the system in agreement?
- Can the position be supported by background theories?
- Does the system do justice to theoretical virtues?
- When comparing input commitments and resulting commitments, is it plausible that we did not abandon the subject?
- Do (at least some of) the resulting commitments have independent credibility?
- Is the resulting position at least as plausible as relevant alternatives?

The answers to these questions will most likely not be a clear yes or no, but rather a matter of degree: a position can be more or less supported by background theories, a system can have a higher or a lower degree of theoretical virtuousness, etc. To answer the question of whether a reflective equilibrium was reached—whether, on balance, the different criteria are met to a sufficient degree—thus also depends, once again, on the pragmatic-epistemic objective; and on whether or not there are better alternatives available. The only thing that should not be traded off in favor of other criteria is consistency—admitting jointly inconsistent claims would undermine the epistemic enterprise (Elgin 1996, 103).

If the answers to these questions are not satisfactory, we could, e.g., retrace some of the steps of the RE process and explore different pathways of adjustments, e.g., choosing to adjust the system in case of a conflict where previously the commitments had been adjusted.

Indeed, there is no guarantee that the method will lead to a justified account, as has for example been brought forward against RE by Kelly and McGrath (2010). But this is why we have the criteria to assess whether such a state was reached, and if so, to what degree. And if such a state was reached, then we have also reached a "fallible, provisional, but reasonable epistemological stance" (Elgin 2014, 255). At any point, our equilibrium can become unbalanced again, so that we will need to continue the process of equilibration. But this does not make the justification via RE useless, or unreasonable.

3.4 Recapitulation: A Methodology of Reflective Equilibrium

This chapter developed a methodological framework for specifying RE as a method which is intended to be applied by one epistemic agent. This methodology was developed with respect to the goal of justifying systematic accounts of a subject matter, for example, in the form of theories or principles. The chapter highlighted methodological and pragmatic decisions that one has to make in order to concretize a method of RE for specific justificatory projects. A lot of these issues are not inherent to reflective equilibrium, but have to be resolved by other research fields that can tell us, e.g., what counts as good evidence, what a strong non-deductive inference is, when testimony is reliable, and so on. Still, these decisions need to be made in order to be able to apply RE as a method. How they are made always depends on the specific project at hand.

The specific project in this book concerns the justification of a precautionary principle as a case study on the applicability of RE. Thus, in the next Chap. 4, I conduct a literature survey on precautionary principles in order to familiarize myself—and my readers—with the subject matter. This survey will also serve as a source for identifying commitments, background elements, and candidates for the system.

Chapter 5 then describes the setup for the case study. It identifies the elements of the initial position, and makes the necessary methodological decisions in order to concretize the RE method for its application to this specific project, that is, it addresses the tasks that have been identified in the present chapter. The case study itself is conducted in Chaps. 6–8.

References

Baumberger C, Brun G (2017) Dimensions of objectual understanding. In: Grimm SR, Baumberger C, Ammon S (eds) Explaining understanding: new perspectives from epistemology and philosophy of science. Routledge, New York, pp 165–189

Beauchamp TL, Childress JF (2013) Principles of biomedical ethics, 7th edn. Oxford University Press, Oxford

Beisbart C, Betz G, Brun G (2021) Making reflective equilibrium precise: a formal model. Ergo 8(15):441–472. https://doi.org/10.3998/ergo.1152

Bonevac D (2004) Reflection without equilibrium. J. Philos. 101(7):363–388

Brun G (2020) Conceptual re-engineering: from explication to reflective equilibrium. Synthese 197(3):925–954. https://doi.org/10.1007/s11229-017-1596-4

Cath Y (2016) Reflective equilibrium. In: Cappelen H, Gendler T, Hawthorne J (eds) The oxford handbook of philosophical methodology. Oxford University Press, Oxford, pp 213–230

de Vries M, van Leeuwen E (2010) Reflective equilibrium and empirical data: third person moral experiences in empirical medical ethics. Bioethics 24(9):490–498. https://doi.org/10.1111/j.1467-8519.2009.01721.x

Elgin CZ (1996) Considered judgment. Princeton University Press, Princeton

Elgin CZ (2014) Non-foundationalist epistemology: holism, coherence, and tenability. In: Steup M, Turri J, Sosa E (eds) Contemporary debates in epistemology, 2nd edn. Contemporary Debates in Philosophy. Wiley Blackwell, Chichester, pp 244–255

Fricker M (2007) Epistemic injustice: power and the ethics of knowing. Oxford University Press, Oxford

Goodman N (1951) The structure of appearance. Harvard University Press, Cambridge

Goodman N (1983) Fact, Fiction, and Forecast, 4th edn. Harvard University Press, Cambridge

Kelly T, McGrath S (2010) Is Reflective equilibrium enough? Philos Perspect 24(1):325–359. https://doi.org/10/bmh3g5

Kemp S, Grace RC (2010) When can information from ordinal scale variables be integrated? Psychol Methods 15(4):398–412. https://doi.org/10.1037/a0021462

Knight C (2017) Reflective equilibrium. In: Blau A (ed) Methods in analytical political theory. Cambridge University Press, Cambridge, pp 46–64

Okasha S (2011) Theory choice and social choice: Kuhn versus arrow. Mind 120(477):83–115. https://doi.org/10.1093/mind/fzr010

Rawls J (1999) A theory of justice. Revised Edition. Belknap Press, Cambridge

Stegenga J (2015) Theory choice and social choice: Okasha versus Sen. Mind 124(493):263–277. https://doi.org/10.1093/mind/fzu180

Tersman F (1993) Reflective equilibrium: an essay in moral epistemology. Lagerblads tryckeri AB, Karlshamn

Van der Burg W, Van Willigenburg T (1998) Introduction. In: Van der Burg W, Van Willigenburg T (eds) Reflective equilibrium: essays in honour of Robert Heeger, library of ethics and applied philosophy, vol 2. Springer, Netherlands, pp 1–25

Van Thiel GJ, Van Delden JJ (2009) The justificatory power of moral experience. J Med Ethics 35(4):234–237

Walden K (2013) In defense of reflective equilibrium. Philos Stud 166(2):243–256. https://doi.org/10.1007/s11098-012-0025-2

Chapter 4
Precautionary Principles

Chapters 2 and 3 introduced the theoretical foundations of reflective equilibrium (RE) and elaborated on how it can be applied in the form of a method. In order to test whether this RE conception can in fact be applied, and to determine what we can learn from such an application, I will conduct a detailed case study in which I demonstrate how RE can be used to justify a precautionary principle. Thus, it is important that we first familiarize ourselves with the subject matter of precaution and precautionary principles.

The basic idea underlying a precautionary principle (PP) is often summarized as "better safe than sorry": even if it is uncertain whether an activity will lead to harm, for example, to the environment or to human health, measures should be taken to prevent harm. This demand is partly motivated by the consequences of regulatory practices of the past: often, chances of harm were disregarded because there was no scientific proof of a causal connection between an activity or substance and chances of harm, for example, between asbestos and lung diseases. When the connection was finally established, it was often too late to prevent severe damage. However, it is highly controversial how the vague intuition behind "better safe than sorry" should be understood as a principle. As a consequence, we find a multitude of interpretations ranging from decision rules over epistemic principles to procedural frameworks. To acknowledge this diversity, it makes sense to speak of precautionary principles (PPs) in the plural. PPs are not without critiques. For example, it has been argued that they are paralyzing, unscientific, or that they promote a culture of irrational fear.

After introducing the main idea and motivation behind precautionary principles in Sect. 4.1, this chapter gives an overview of different PP interpretations according to their function (Sect. 4.2). It then describes the main lines of arguments that have been presented in favor of PPs (Sect. 4.3). Section 4.4 presents the most frequent and most important objections that PPs face, along with possible rejoinders. Lastly, Sect. 4.5 recapitulates the main points from the perspective of the case study for reflective equilibrium. One important function of this survey with respect to the case

T. Rechnitzer, *Applying Reflective Equilibrium*, Logic, Argumentation & Reasoning 27, https://doi.org/10.1007/978-3-031-04333-8_4

study for reflective equilibrium is to help us in identifying plausible candidates for the system. In particular, I take it that in order to count as a successful clarification and formulation of a PP, such a proposed principle at the minimum has to:

- Name clear conditions under which the PP applies and the requirements that follow from it;
- And since it expresses a normative claim, it has to be stated on which grounds the principle is justified.

In addition to identifying relevant candidates for the system, this survey will help us (i) to make informed commitments concerning the subject matter, (ii) to identify relevant background theories, and (iii) to be aware of challenges that the target system should be able to address.

4.1 The Idea of Precaution and Precautionary Principles

We can identify three main motivations behind the postulation of a PP. First, it stems from a deep dissatisfaction with how decisions were made in the past: often, early warnings have been disregarded, leading to significant damage which could have been avoided by timely precautionary action (Harremoës et al. 2001). This motivation for a PP rests on some sort of "inductive evidence" that we should reform (or maybe even replace) our current practices of risk regulation, demanding that uncertainty must not be a reason for inaction (John 2007).

Second, it expresses specific moral concerns, usually pertaining to the environment, human health, and/or future generations. This second motivation is often related to the call for sustainability and sustainable development in order to not destroy important resources for short-term gains, but to leave future generations with an intact environment.

Third, PPs are discussed as principles of rational choice under conditions of uncertainty and/or ignorance. Typically, rational decision theory is well suited for situations where we know the possible outcomes of our actions and can assign probabilities to them (a situation of "risk" in the decision-theoretic sense). However, the situation is different for decision-theoretic uncertainty (where we know the possible outcomes, but cannot assign any, or at least no meaningful and precise, probabilities to them) or decision-theoretic ignorance (where we do not know the complete set of possible outcomes). Although there are several suggestions for decision rules under these circumstances, it is far from clear what is the most rational way to decide when we are lacking important information and the stakes are high. PPs are one proposal to fill this gap.

Although they are often asserted individually, these motivations also complement each other: if, as following from the first motivation, uncertainty is not allowed to be a reason for inaction, then we need some guidance for how to decide under such circumstances, for example, in the form of a decision principle. And in many cases, it is the second motivation—concerns for the environment or human health—which makes the demand for precautionary action before obtaining scientific certainty especially pressing.

Many existing official documents cite the demand for precaution. One often-quoted example for a PP is principle 15 of the Rio Declaration on Environment and Development, a result of the United Nations Conference on Environment and Development (UNCED) 1992. It refers to a "precautionary approach":

Rio PP In order to protect the environment, the precautionary approach shall be widely applied by states according to their capabilities. Where there are threats of serious or irreversible damage, lack of full scientific certainty shall not be used as a reason for postponing cost-effective measures to prevent environmental degradation. (United Nations Conference on Environment and Development 1992, Principle 15)

Another prominent example is the formulation that resulted from the Wingspread Conference on the Precautionary Principle 1998, where around 35 scientists, lawyers, policy makers and environmentalists from the United States, Canada, and Europe met to define a PP:

Wingspread PP When an activity raises threats of harm to human health or the environment, precautionary measures should be taken even if some cause-and-effect relationships are not fully established scientifically. In this context the proponent of an activity, rather than the public, should bear the burden of proof. The process of applying the precautionary principle must be open, informed and democratic and must include potentially affected parties. It must also involve an examination of the full range of alternatives, including no action. (Science & Environmental Health Network (SEHN) 1998)

Both formulations are often cited as paradigmatic examples of PPs. Although they both mention uncertain threats and measures to prevent them, they also differ in important points, for example their strength: the Rio PP makes a weaker claim, stating that uncertainty is not a reason for inaction, whereas the Wingspread PP puts more emphasis on the fact that measures should be taken. They both give rise to a variety of questions: what counts as "serious or irreversible damage"? What does "(lack of) scientific certainty" mean? How plausible does a threat have to be in order to warrant precaution? What counts as precautionary measures? Additionally, PPs face many criticisms, like being too vague to be action-guiding, paralyzing the decision-process, or being anti-scientific and promoting a culture of irrational fear.

Thus, inspired by these regulatory principles in official documents, a lively debate has developed around how PPs should be interpreted in order to arrive at a version applicable in practical decision-making. This has resulted in a multitude of PP proposals that are formulated and defended (or criticized) in different theoretical and practical contexts. Most of the existing PP formulations share the elements of uncertainty, harm, and (precautionary) action. Different ways of spelling out these elements result in different PPs (Manson 2002; Sandin 1999). For example, they can vary in how serious harm has to be in order to trigger precaution, or which amount of evidence is needed. Additionally, PP interpretations differ with respect to the function they are intended to fulfill. They are typically classified based on some combination of the following categories according to their function (Munthe 2011; Sandin 2007, 2009; Steel 2015):

- *Action-guiding* principles tell us which course of action to choose given specific circumstances;
- (sets of) *epistemic* principles tell us what we should reasonably believe under conditions of uncertainty;
- *procedural* principles express requirements for decision-making, and tell us how we should choose a course of action.

These categories can overlap, for example, when action- or decision-guiding principles come with at least some indication for how they should be applied. Some interpretations explicitly aim at integrating the different functions, and warrant their own category:

- *Integrated* PP interpretations: Approaches that integrate action-guiding, epistemic, and procedural elements associated with PPs. Consequently, they tell us which course of action should be chosen through which procedure, and on what epistemic basis.

4.2 Interpretations of Precautionary Principles

This section gives an overview of different interpretations of precautionary principles according to the functions described above. For the case study on the applicability of reflective equilibrium, this will help us to identify candidates for the system.

4.2.1 Action-Guiding Interpretations

Action-guiding PPs are often seen as on a par with decision rules from rational decision theory. On the one hand, authors formalize PPs by using decision rules already established in decision theory, like maximin. On the other hand, they formulate new principles. While not necessarily located within the framework of decision theory, those are intended to work at the same level. Understood as principles of risk management, they are supposed to help to determine a course of action given our knowledge and our values.

4.2.1.1 Decision Rules

The terms used for decision-theoretic categories of non-certainty differ. I will use them as follows: *decision-theoretic risk* denotes situations in which we know the possible outcomes of actions and can assign probabilities to them. *Decision-theoretic uncertainty* refers to situations in which we know the possible outcomes, but either no or only partial or imprecise probability information is available (Hansson 2005a, 27). When we don't even know the full set of possible outcomes, we have a situation of *decision-theoretic ignorance*. When formulated as decision rules, the "(scientific) uncertainty" component of PPs is often spelled out as decision-theoretic uncertainty.

Maximin

The idea to operationalize a PP with the maximin decision rule occurred early within the debate and is therefore often associated with PPs (e.g., Aldred 2013; Gardiner 2006; Hansson 1997; Sunstein 2005b).

In order to be able to apply the maximin rule, we have to know the possible outcomes of our actions and be able to at least rank them on an ordinal scale (meaning that for each outcome, we can tell whether it is better, worse, or equally good than each other possible outcome). It then tells us to select the option with the best worst case in order to "maximize the minimum". Thus, the maximin rule seems like a promising candidate for a PP. It pays special attention to the prevention of threats, and is applicable under conditions of uncertainty. However, as has repeatedly been pointed out, maximin is not a plausible rule of choice in general. Consider the decision matrix in Table 4.1.

Maximin selects Alternative_1. This seems excessively risk-averse because the best case in Alternative_2 is much better, and the worst case is only slightly worse (as long as we assume (a) that the utilities in this example are cardinal utilities, and (b) that there is not some kind of relevant threshold passed). If we knew that the probability for Scenario_1 is 0.99 and the probability for Scenario_2 only 0.01—then it would arguably be absurd to apply maximin. Proponents of interpreting a PP with maximin thus have stressed that it needs be qualified by some additional criteria in order to provide a plausible PP interpretation.

Table 4.1 Simplified decision-matrix with two alternative courses of action

	Scenario_1	Scenario_2
Alternative_1	7	6
Alternative_2	15	5

The most prominent example is Gardiner (2006), who draws on criteria suggested by Rawls to determine the conditions under which the application of maximin is plausible:

1. Knowledge of likelihoods for the possible outcomes of the actions is impossible or at best extremely insecure;
2. the decision-makers care relatively little for potential gains that might be made above the minimum that can be guaranteed by the maximin approach;
3. the alternatives that will be rejected by maximin have unacceptable outcomes; and
4. the outcomes considered are in some adequate sense "realistic", that is, only credible threats should be considered.

Condition (3) makes it clear that the guaranteed minimum (condition 2) needs to be acceptable to the decision-makers (see also Rawls 2001, 98). What it means that gains above the guaranteed minimum are relatively little cared for (condition 2) has been spelled out by Aldred (2013) in terms of incommensurability between outcome values, that is, that some outcomes are so bad that they cannot be outweighed by potential gains. It is thus better to choose an option that promises only little gains but guarantees that the extremely bad outcome can't materialize.

Gardiner argues that a maximin rule that is qualified by these criteria fits well with some core cases where we agree that precaution is necessary, and calls it the "Rawlsian Core Precautionary Principle (RCPP)". He cites the purchase of insurance as an everyday example where his RCPP fits well with our intuitive judgments and where precaution seems already justified on its own. According to Gardiner, it also fits well with often-cited paradigmatic cases for precaution like climate change: the controversy concerning whether we should take precautions in the climate case is not a debate around the right interpretation of the RCPP but rather about whether the conditions for its application are fulfilled—for example, which outcomes are unacceptable (Gardiner 2006, 56).

Minimax Regret

Another decision rule that is occasionally discussed in the context of PPs is the minimax regret rule[1] (Chisholm and Clarke 1993; Iverson and Perrings 2012). The rule selects the course of action where under each alternative scenario, the maximal regret is the smallest. Chisholm and Clarke (1993) strongly support the minimax regret rule, arguing that it is better suited for PP than maximin, since it gives some weight to benefits foregone. They also show that even if it is uncertain whether or not precautionary measures will be effective, minimax regret still recommends them as long as the expected damage from not applying PP is

[1] For an explanation of minimax regret, and a short comparison with the maximin rule, see Hansson (2005a, 61–62).

large enough. They advocate so-called "dual purpose" policies, where precautionary measures still have other positive effects, even if they do not fulfill their main purpose (e.g., measures which are aimed at abating global climate change, but at the same time have direct positive effects on local environmental problems). Thus, it seems that by applying minimax regret to constructed examples, they want to support a specific PP interpretation, and do not directly propose minimax regret as a PP. However, besides citing a PP from the Bergen Ministerial Declaration (1990)[2] it remains unclear how their argumentation relates to this PP, or if they have their own interpretation in mind.

4.2.1.2 Context-Sensitive Principles

Other interpretations of PPs as action-guiding *principles* differ from stand-alone if-this-then-that decision rules. They stress that principles have to be interpreted and concretized depending on the specific context (Fisher 2002; Randall 2011).

A Virtue Principle

Sandin (2009) argues that one can reinterpret a PP as an action-guiding principle not by reference to decision theory, but by using cautiousness as a virtue. He formulates an action-guiding virtue principle of precaution (VPP):

VPP Perform those, and only those, actions that a cautious agent would perform in the circumstances. (Sandin 2009, 98)

Although virtue principles are commonly criticized as not being action-guiding, Sandin argues that understanding a PP in this way actually makes it more action-guiding. "Cautious" is interpreted as a virtue term that refers to a property of an agent, like "courageous" or "honest". Sandin states that it is often possible to identify what the virtuous agent would do: Either because it is obvious, or because at least some agreement can be reached. Even the uncertain cases VPP deals with belong to classes of situations where we have experience with, for example, failed regulations of the past, and therefore can assess what the cautious agent would (not) have done and extrapolate from that to other cases (Sandin 2009, 99). According to Sandin, interpreting a PP as a virtue principle will avoid both objections of extremism and paralysis. It is unlikely that the virtuous agent will choose courses of action which will, in the long run, have overall negative effects or are self-refuting (like "ban activity a and do not ban activity a!"). However, even if one accepts that it makes sense to interpret "cautious" as a virtue, "the circumstances" under which one should choose the course of action that the cautious agent would choose are

[2] "Where there are threats of serious or irreversible damage, lack of full scientific certainty should not be used as a reason for postponing measures to prevent environmental degradation."

not specified in the VPP as it is formulated by Sandin. This makes it an incomplete proposal.

Reasonableness and Plausibility

Another important example is the PP interpretation by Resnik (2003, 2004), who defends a PP as an alternative to maximin and other strategies for decision-making in situations where we lack the type of empirical evidence that one would need for a form of risk management that uses probabilities obtained from risk assessment. His PP interpretation, which we can call the "reasonable measures precautionary principle (RMPP)", reads as follows:

RMPP One should take reasonable measures to prevent or mitigate threats that are plausible and serious.

The seriousness of a threat relates to its potential for harm, as well as to whether the possible damage is seen as reversible or not (Resnik 2004, 289). Resnik emphasizes that reasonableness is a highly pragmatic and situation-specific concept. He offers some criteria for reasonable responses that are neither exhaustive nor necessary: they should be effective, proportional to the nature of the threat, take a realistic attitude toward the threat, be cost-effective, and be applied consistently (Resnik 2003, 341–42). Lastly, that threats have to be credible means that there have to be scientific arguments for the plausibility of a hypothesis. These can be based on epistemic and/or pragmatic criteria, including for example coherence, explanatory power, analogy, precedence, precision, or simplicity. Resnik stresses that a threat being plausible is not the same as a threat being even minimally probable: we might accept threats as plausible that we think to be all but impossible to come to fruition (Resnik 2003, 342).

This shows that the question of when a threat should count as plausible enough to warrant precautionary measures is very important for the application of an action-guiding PP. Consequently, such PPs are often very sensitive to how a problem is framed. Some authors have taken these aspects—the weighing of evidence and the description of the decision problem—to be central points of PPs, and interpreted them as epistemic principles, that is, principles at the level of risk assessment.

4.2.2 *Epistemic Interpretations*

Authors who defend an epistemic PP interpretation argue that we should accept that PPs are not principles that can guide our actions, but that this is neither a problem nor against their spirit. Instead of telling us how to act when facing uncertain threats of harm, they propose that PPs tell us something about how we should perceive these threats, and what we should take as a basis for our actions, for example, by relaxing the standard for the amount of evidence required to take action.

4.2.2.1 Standards of Evidence

One interpretation of an epistemic PP is to give more weight to evidence suggesting a causal link between an activity and threats of serious and irreversible harm than one gives to evidence suggesting less dangerous, or beneficial, effects. This could mean to assign a higher probability for an effect to occur than one would in other circumstances based on the same evidence. Arguably, the underlying idea of this PP can be traced back to the German philosopher Hans Jonas, who proposed a "heuristic of fear", that is, to give more weight to pessimistic forecasts than to optimistic ones (Jonas 1979). However, this PP interpretation has been criticized on the basis that it systematically discounts evidence pointing in one direction, but not in the other. This could lead to distorted beliefs about the world in the long run, being detrimental to our epistemic and scientific progress and eventually doing more harm than good (Harris and Holm 2002).

However, other authors point out that we might have to distinguish between "regulatory science" and "normal science". Different epistemic standards are appropriate for the two contexts since they have different aims: in normal science, we are searching for truth; in regulatory science, we are primarily interested in reducing risk and avoiding harm (John 2010). Accordingly, Peterson (2007b) refers in his epistemic PP interpretation only to decision makers—not scientists—who find themselves in situations involving risk or uncertainty. He argues that in such cases, decision-makers should strive to acquire beliefs that are likely to protect human health, and that it is less important whether they are also likely to be true. One principle that has been promoted in order to capture this idea is the preference for false positives, that is, for type I errors over type II errors.

4.2.2.2 Type I and Type II Errors

Is it worse to falsely believe that there is a relationship between two classes of events, which does not exist (false positives), or to fail to assert such a relationship, when it in fact exists (false negatives)? For example, would you prefer virus software on your computer which classifies a harmless program as a virus (false positive) or rather one that misses a malicious program (false negative)? Statistical hypotheses testing tests the so-called null-hypothesis, which is the default view that there is no relationship between two classes of events, or groups. Rejecting a true null hypothesis is called a type I error, whereas failing to reject a false null hypothesis is a type II error. Which type of possible error should we try to minimize, if we cannot minimize both at once?

In (normal) science, it is more highly valued not to include false assertions into the body of knowledge, since these would distort it in the long term. Thus, the default assumption—the null hypothesis—is that there is no connection between two classes of events, and typically statistical procedures are used that minimize type I errors (false positives) even if this might mean that an existing connection is missed (at least at first, or for a long time) (John 2010). To believe that a certain existing deterministic or probabilistic connection between two classes of events does not exist might slow down the scientific progress in normal science aiming at truth. However, in regulatory contexts it might be disastrous to believe falsely that a substance is safe when it is not. Consequently, a prominent interpretation of an epistemic PP takes it to entail a preference for type I errors over type II errors in regulatory contexts (see for example John 2010; Lemons et al. 1997; Peterson 2007b).

Merely favoring one type of error over another might not be enough. It has been argued that the underlying methodology of either rejecting or accepting hypotheses does not sufficiently allow for identifying and tracking uncertainties. If a PP is understood as a principle that relaxes the standard for the amount of evidence required to take action, then a new epistemology might be needed: one that allows an integrating of the uncertainty about the causal connection between, for example, a drug and a harm, in the decision (Osimani 2013).

4.2.2.3 Precautionary Defaults

The use of precautionary regulatory defaults is one proposal for how to deal with having to make regulatory decisions in the face of insufficient information (Sandin and Hansson 2002; Sandin et al. 2004). In regulatory contexts, there are often situations in which a decision has to be made on how to treat a potentially harmful substance that also has some (potential) benefits. Unlike in normal science, it is not possible to wait and collect further evidence before a decision is made. The substance has to be treated one way or another while waiting for further evidence. Thus, it has been suggested that we should use regulatory defaults, i.e., assumptions that are used in the absence of adequate information and that should be replaced if such information were obtained. They should be precautionary defaults by building in special margins of safety in order to make sure that the environment and human health get sufficient protection. One example is the use of uncertainty factors in toxicology. Such uncertainty factors play a role in estimating reference doses which are acceptable for humans by dividing a level of exposure found acceptable in animal experiments by a number (usually 100) (Steel 2011, 356). This takes into account that there are significant uncertainties, for example, in extrapolating the results from animals to humans. Such defaults are a way to handle uncertain threats. Nevertheless, they should not be confused with actual judgments about what properties a particular substance has (Sandin et al. 2004, 5). Consequently, an epistemic PP does not have to be understood as a belief-guiding principle, but as saying something about which methods for risk assessment are legitimate, for example, for quantifying uncertainties (Steel 2011). According to this view, precautionary defaults like uncertainty factors in toxicology are methodological implications of a PP that allow it to be applied in a scientifically sound way while protecting human health and the environment.

Given this, it might be misleading to interpret a PP as a purely epistemic principle, if it is not guiding our beliefs but telling us what assumptions to accept, i.e., telling us to act *as if* certain things were true, as long as we do not have more information. Thus, it has been argued that a PP is better interpreted as a procedural requirement, or as a principle that imposes several such procedural requirements (Sandin 2007, 103–04).

4.2.3 Procedural Interpretations

The thought, then, is that we should shift our attention when interpreting PPs: from the question of what action to take to the question of what is the best way to reach decisions.

4.2.3.1 Argumentative, or "Meta" PPs

Argumentative PPs are procedural principles specifying what kinds of arguments are admissible in decision-making (Sandin et al. 2002). They are different from prescriptive, or action-guiding, PPs in that they do not directly prescribe actions that should be taken. Take principle 15 of the Rio Declaration on Environment and Development. On one interpretation, it states that arguments for inaction which are based solely on the ground that we are lacking full scientific certainty, are not acceptable arguments in the decision-making procedure:

Rio PP In order to protect the environment, the precautionary approach shall be widely applied by states according to their capabilities. Where there are threats of serious or irreversible damage, lack of full scientific certainty shall not be used as a reason for postponing cost-effective measures to prevent environmental degradation. (United Nations Conference on Environment and Development 1992, Principle 15)

Such an argumentative PP is seen as a meta-rule that places real constraints on what types of decision rules should be used: for example, by entailing that decision-procedures should be used that are applicable under conditions of uncertainty, it recommends against some of the traditional approaches in risk regulation like cost-benefit analysis (Steel 2015). Similarly, it has been proposed that the idea behind PPs is best interpreted as a general norm that demands a fundamental shift in our way of risk regulation, based on an obligation to learn from regulatory mistakes of the past (Whiteside 2006).

4.2.3.2 Transformative Decision Rules

Similar to argumentative principles, an interpretation of a PP as a transformative decision rule does not tell us which action should be taken, but it puts constraints on which actions can be considered as valid options. Informally, a transformative decision rule is defined as a decision rule that takes one decision problem as input, and yields a new decision problem as output (Sandin 2004, 7). For example, the following formulation of a PP as a transformative decision rule (TPP) has been proposed by Peterson (2003):

TPP If there is a non-zero probability that the outcome of an alternative act is very low, i.e. below some constant c, then this act should be removed from the decision-maker's list of options.

Thus, the TPP excludes courses of action that could lead, for example, to catastrophic outcomes, from the options available to the decision maker. However, it does not tell us which of the remaining options should be chosen.

4.2.3.3 Reversing the Burden of Proof

The requirement of reversal of burden of proof is one of the most prominent specific procedural requirements that are cited in connection with PPs. For example, in the influential communication on the PP from the Wingspread Conference on the Precautionary Principle (1998), it is stated, "the proponent of an activity, rather than the public bears the burden of proof".

One common misconception is that the proponent of a potentially dangerous activity would have to prove with absolute certainty that the activity is safe. This gave rise to the objection that PPs are too demanding, and therefore would bring progress to a halt (Harris and Holm 2002). However, the idea is rather that we have to change our approach to regulatory policy: proponents of an activity have to prove to a certain threshold that it is safe in order to employ it, instead of the opponents having to prove to a certain threshold that it is harmful in order to ban it.

Thus, whether or not the situation is one in which the burden of proof is reversed depends on the status quo. Instead of speaking of shifting the burden of proof, it seems more sensible to ask what has to be proven, and who has to provide what kind of evidence for it. The important point that then remains to be clarified is what standards of proof are accepted.

An alternative proposal to shifting the burden of proof is that both regulators and proponents of an activity should share it (Arcuri 2007): if opponents want to regulate an activity, they should at least provide some evidence that the activity might lead to serious or irreversible harm, even though we are lacking evidence to prove it with certainty. Proponents, on the other hand, should provide some information about the activity in order to get it approved. Who has the burden of proof can thus play an important role in the production of information: if proponents have to show (to a specific standard) that their activity is safe, this generates an incentive to gather information about the activity, whereas in the other case—"safe until proven otherwise"—they might deliberately refrain from this (Arcuri 2007, 15).

4.2.3.4 Procedures for Determining Precautionary Measures

Interpreted in a procedural way, a PP puts constraints on how a problem should be described or how a decision should be made. It does not dictate a specific decision or action. This is in line with one interpretation of what it means to be a principle as opposed to a rule. While rules specify precise consequences that follow automatically when certain conditions are met, principles are understood as guidelines whose interpretation will depend on specific contexts (Arcuri 2007; Fisher 2002).

Developing a procedural precautionary framework that integrates different procedural requirements is a way to enable the context-dependent specification and implementation of such a PP. One example is Tickner's (2001) "precautionary assessment" framework, which consists of six steps that are supposed to guide decision-making as a heuristic device. The first five steps—(1) Problem Scoping, (2) Participant Analysis, (3) Burden/Responsibility Allocation Analysis, (4) Environment and Health Impact Analysis, and (5) Alternatives Assessment—serve to describe the problem, identify stakeholders, and to assess possible consequences as well as available alternatives. In the final step, (6) Precautionary Action Analysis, the appropriate precautionary measure(s) are determined based on the results from the other steps. These decisions are not permanent, but should be part of a continuous process of increasing understanding and reducing overall impacts.

That the components are clarified on a case-by-case basis is a big advantage of such procedural implementations of PPs. It avoids an oversimplification of the decision process and takes the complexity of decisions under uncertainty into account. However, they are criticized for losing the "principle" part of PPs: for example, Aldred (2013) argues that procedural requirements form a heterogeneous category. A procedural PP would soon dissolve beyond recognition because it is intermingled with other (rational, legal, moral, and so forth) principles and rules. As an answer, some authors try to preserve the "principle" in PPs, while also taking into account procedural as well as epistemic elements.

4.2.4 Integrated Interpretations

We can find two main strategies for formulating a PP that is still identifiable as an action-guiding principle while integrating procedural as well as epistemic considerations: either (1) developing particular principles that are specific to a certain context, and accompanied by a procedural framework for this context; or (2) describing the structure and the main elements of a PP plus identifying criteria for adjusting those elements on a case-by-case basis.

4.2.4.1 Particular Principles for Specific Contexts

It has been argued that the general talk of "the" PP should be abandoned in favor of formulating distinct precautionary principles (Hartzell-Nichols 2013). This strategy aims to arrive at action-guiding and coherent principles by formulating particular PPs that apply to a narrow range of threats and express a specific obligation. One example is the "Catastrophic Harm PP (CHPP)" of Hartzell-Nichols (2012, 2017), which is restricted to catastrophic threats. It consists of eight conditions that specify when precautionary measures have to be taken, spelling out (a) what counts as a catastrophe, (b) the knowledge requirements for taking precaution, and (c) criteria for appropriate precautionary measures. The CHPP is accompanied by a "Catastrophic Precautionary Decision-Making Framework" which guides the assessment of threats in order to decide whether they meet the CHPP's criteria, and guides decision-makers in determining what precautionary measures should be taken against a particular threat of catastrophe. This framework lists key considerations and steps that should be performed when applying the CHPP, for example, drawing on all available sources of information, assessing likelihoods of potential harmful outcomes under different scenarios, identifying all available courses of precautionary action and their effectiveness, and identifying specific actors who should be held responsible for taking the prescribed precautionary measures.

4.2.4.2 An Adjustable Principle with Procedural Instructions

Identifying the main elements of a PP and accompanying them with rules for adjusting them on a case-by-case basis is another strategy to preserve the idea of a precautionary principle while avoiding both inconsistency as well as vagueness. It has been shown that as diverse as PP formulations are, they typically share the elements of uncertainty, harm, and (precautionary) action (Manson 2002; Sandin 1999). By explicating these concepts and, most importantly, by defining criteria for how they should be adjusted with respect to each other, some authors obtain a substantial PP that can be adjusted on a case-by-case basis without becoming arbitrary.

One example is the PP that Randall (2011) develops in the context of an in-depth analysis of traditional, or as he calls it, ordinary risk management (ORM). Randall identifies the following "general conceptual form of PP":

> If there is evidence stronger than **E** that an activity raises a threat more serious than **T**, we should invoke a remedy more potent than **R**.

Threat, **T**, is explicated as chance of harm, meaning that threats are assessed and compared according to their magnitude and likelihood. Our knowledge of outcomes and likelihoods is explicated with the concept of *evidence*, **E**, referring to uncertainty in the sense of our incomplete knowledge about the world. The precautionary response is conceptualized as *remedy*, **R**, which covers a wide range of responses such as averting the threat, remediating its damage, mitigating harm, and adapting to changed conditions after other remedies have been exhausted. Remedies should fulfill a double function, (1) providing protection from a plausible threat, while at the same time (2) generating additional evidence about the nature of the threat and the effectiveness of various remedial actions. The main relations between the three elements are that the higher the likelihood that the remedy-process will generate more evidence, the smaller is the threat-standard and the lower is the evidence-standard that should be required before invoking the remedy even if we have concerns about its effectiveness (Randall 2011, 167).

Having clarified the concepts used in the ETR-framework, Randall specifies them in order to formulate a PP that accounts for the weaknesses of ORM:

> Credible scientific evidence of plausible threat of disproportionate and (mostly but not always) asymmetric harm calls for avoidance and remediation measures beyond those recommended by ordinary risk management. (Randall 2011, 186)

He then goes on to integrate this PP and ORM together into an integrated risk-management framework. Randall makes sure to stress that a PP cannot determine the decision-process on its own. As a moral principle, it has to be weighed against other moral, political, economic, and legal considerations. Thus, he also calls for the development of a procedural framework to ensure that its substantive normative commitments will be implemented on the ground (Randall 2011, 207).

Steel (2013, 2015) develops a comprehensive PP interpretation which is intended to be "a procedural requirement, a decision rule, and an epistemic rule" (Steel 2015, 14). Referring to the Rio Declaration, Steel argues that such a formulation of a PP states that our decision-process should be structured differently, namely that decision rules should be used that can be applied in an informative way under uncertainty. However, he does not take this procedural element to be the whole PP, but interprets it as a "meta"-rule which guides the application and specification of the precautionary "tripod" of threat, uncertainty, and precautionary action. More specifically, Steel's proposed PP consists of three core elements:

The Meta Precautionary Principle (MPP) Uncertainty must not be a reason for inaction in the face of serious threats.

The Precautionary Tripod The elements that have to be specified in order to obtain an action-guiding version of the precautionary principle, namely: If there is a threat that meets the *harm condition* under a given *knowledge condition* then a *recommended precaution* should be taken.

Proportionality Demands that the elements of the Precautionary Tripod are adjusted proportionally to each other, understood as *Consistency*: The recommended precaution must not be recommended against by the same PP version, and *Efficiency*: Among those precautionary measures that can be consistently recommended by a PP version, the least costly one should be chosen.

An application of this PP requires selecting what Steel calls a "relevant version of PP", that is, a specific instance of the Precautionary Tripod that meets the constraints from both MPP and Proportionality. To obtain such a version, Steel (2015, 30) proposes the following strategy: (1) select a desired safety target and define the harm condition as a failure to meet this target, (2) select the least stringent knowledge condition that results in a consistently applicable version of PP given the harm condition. To comply with the MPP, uncertainty must neither render the PP version inapplicable nor lead to continual delay in taking measures to prevent harm.

Thus, Steel's PP proposal guides decision-makers both in formulating the appropriate PP version as well as in its application. The process of formulating the particular version already deals with many questions such as how evidence should be assessed, who has to prove what, to what kind of threats we should react, and what the appropriate precautionary measures would be. Arguably, this PP can thereby be action-guiding, since it helps to select specific measures, without being a rigid prescriptive rule that is not suited for decisions under uncertainty.

Additionally, proposals like the ones of Randall and Steel have the advantage that they are not rigidly tied to a specific category of decision-theoretic non-certainty, i.e., decision-theoretic risk, uncertainty, or ignorance. They can be adjusted with respect to varying degrees of knowledge and available evidence, taking into account that we typically have some imprecise or vague sense of how likely various outcomes are, but not enough of a sense to assign meaningful precise probabilities to the outcomes. While these situations do not amount to decision-theoretic risk, they nonetheless include more information than what is often taken to be available in decision-theoretic uncertainty. Arguably, this better corresponds to the notion of "scientific uncertainty" than to equate the latter with decision-theoretic uncertainty (see Steel 2015, chapter 4).

4.3 Justifications for Precautionary Principles

This section surveys different normative backgrounds that have been used to defend a PP. In the context of the case study for reflective equilibrium, this will help us to identify relevant background theories. I start by addressing arguments that can be located in the framework of practical rationality, before moving to substantial moral justifications for precautions.

4.3.1 Practical Rationality

When PPs are proposed as principles of practical rationality, they are typically seen as principles of risk regulation. This includes, but is not reduced to, rational choice theory. When we examine the justifications for PPs in this context, we have to do this against the background of established risk-regulation practices. We can identify a rather standardized approach to the assessment and management of risks, which Randall (2011, 43) calls "ordinary risk management (ORM)".

4.3.1.1 Ordinary Risk Management

Although there are different understandings of ORM, we can identify a rather robust "core" of two main parts. First, a scientific risk assessment is conducted, where potential outcomes are identified and their extent and likelihood estimated (Randall 2011, 43–46). Typically, risk assessment is understood as a quantitative endeavor, expressing numerical results (Zander 2010, 17). Second, on the basis of the data obtained from the risk assessment, the risk-management phase takes place. Here, alternative regulatory courses of action as response to the scientifically estimated risks are discussed, and a choice is made between them. While the risk-assessment phase should be as objective and value-free as possible, the decisions that take place in the risk-management phase should be, although informed by science, based on the values and interests of the parties involved. In ORM, cost-benefit analysis (CBA) is a powerful and widely used tool for making these decisions in the risk-management phase. To conduct a CBA, the results from the risk assessment, i.e., what outcomes are possible under which course of action, are evaluated according to the willingness to pay (WTP) or willingness to accept compensation (WTA) of individuals in order to estimate the benefits and costs of different courses of action. That means that non-economic values, like human lives or environmental preservation, are monetized in order to be comparable on a common ratio scale. Since we rarely if ever find ourselves facing cases of certainty, where each course of action has exactly one outcome which will materialize if we choose it, these so-reached utilities are then probability-weighed and added up in order to arrive at the expected utility of the different courses of action. On this basis, it is possible to calculate which regulatory actions have the highest expected net benefits (Randall 2011, 47), i.e., to apply the principle of maximizing expected utility (MEU) and to choose the option with the highest expected utility. CBA is seen as a tool that enables decision-makers to rationally compare costs and benefits, helping them to come to an informed decision (Zander 2010, 4).

In the context of ORM, we can distinguish two main lines of argumentation for PPs: on the one hand, authors argue that PPs are rational by trying to show that they gain support from ORM. On the other hand, authors argue that ORM itself is problematic in some aspects, and propose PPs as a supplement or alternative to it. In both cases, we find justifications for PPs as decision rules for risk management as well as principles that pertain to the risk-assessment stage and are concerned with problem-framing (this includes epistemic and value-related questions).

4.3.1.2 PPs in the Framework of Ordinary Risk Management

To begin, here are some ways in which people propose to locate and defend PPs within ORM.

Expected Utility

Some authors claim that as long as we can assign probabilities to the various outcomes, that is, as long as we are in a situation of decision-theoretic risk, precaution is already "built in" into ORM (Chisholm and Clarke 1993; Gardiner 2006; Sunstein 2007). The argument is roughly that no additional PP is necessary because expected utility theory in combination with the assumption of decreasing marginal utility allows for risk aversion by placing greater weight on the disutility of large damages. Not to choose options with possibly catastrophic outcomes, even if they only have a small probability, would thus be recommended by the principle of maximizing expected utility (MEU) as a consequence of their large disutility.

This argumentation does not go unchallenged, as the next subsection shows. Additionally, MEU itself is not uncontroversial (see Buchak 2013). Still, even if we accept it, we cannot use MEU under conditions of decision-theoretic uncertainty, since it relies on probability information. Consequently, authors have proposed PPs for decisions under uncertainty in order to fill this "gap" in the ORM framework. They argue that under decision-theoretic uncertainty, it is rational to be risk-averse, and try to demonstrate this with arguments based on rational choice theory. However, it is not always clear if the discussed decision rule is used to justify a—somehow—already formulated PP, or if the decision rule is proposed as a PP itself.

Maximin and Minimax Regret

Both the maximin rule—selecting the course of action with the best worst case—and the minimax regret rule—selecting the course of action where under each possible scenario, the maximal regret is the smallest—have been proposed and discussed as possible formalizations of a PP within the ORM framework. It has been argued that maximin captures the underlying intuitions of PPs (namely, that the worst should be avoided) and that it yields rational decisions in relevant cases (Hansson 1997). Although the rationality of maximin is contested (Bognar 2011; Harsanyi 1975), it is argued that we can qualify it with criteria to single out the cases in which it can—and should—rationally be applied (Gardiner 2006). This is done by showing that a so-qualified maximin rule fits with paradigm cases of precaution and commonsense decisions that we make, arguing that it is plausible to adopt it also for further cases.

Chisholm and Clarke (1993) argue that the minimax regret rule leads to the prevention of uncertain harm in line with the basic idea of a PP, while also giving some weight to forgone benefits. Against minimax regret and in favor of maximin, Hansson (1997, 297) argues that, firstly, minimax regret presupposes more information, since we need to be able to assign numerical utilities to outcomes. Secondly, he uses a specific example to show that minimax regret and maximin can lead to conflicting recommendations. According to Hansson, the recommendation made by maximin expresses a higher degree of precaution.

Quasi-Option Value

Irreversible harm is mentioned in many PP formulations, for example in the Rio Declaration (see Sect. 4.1). One proposal to justify why "irreversibility" justifies precautions refers to the concept of "(quasi-)option value" (Chisholm and Clarke 1993; Sunstein 2005a, 2009) which was first introduced by Arrow and Fisher (1974). They show that when regulators are confronted with decision problems where they are (a) uncertain about the outcomes of the options, but there are (b) chances for resolving or reducing these uncertainties in the future, and (c) one or more of the options might entail irreversible outcomes, then they should attach an extra value, that is, an option-value to the reversible options. This takes into account the value of the options that choosing an alternative with irreversible outcome would foreclose. To illustrate this, think of the logging of (a part of) the rain forest: it is a very complex ecosystem, which we could use in many ways. But once it is clear-cut, it is effectively impossible to restore to its original state. By choosing the option to cut it down, all options to use the rain forest in any other way would practically be lost forever. As Chisholm and Clarke (1993, 115) point out, irreversibility might sometimes be associated with not taking actions now: not mitigating greenhouse gas (GHG) emissions means that more and more GHG aggregate in the atmosphere, where they stay for a century or more. They argue that introducing the concept of quasi-option value supports the application of a PP even if decision makers are not risk-averse.

4.3.1.3 Reforming Ordinary Risk Management

After reviewing attempts to justify a PP in the ORM framework, without challenging the framework itself, let us now examine justifications for PPs that are partially based on criticisms of ORM.

Deficits of ORM

As a first point, ORM as a regulatory practice tends toward oversimplification that neglects uncertainty and imprecision, leading to irrational and harmful decisions. This is seen as a systematic deficit of ORM itself, not only of its users (Randall 2011, 77), and not only as a problem under decision-theoretic uncertainty, that is, situations where no reliable probabilities are available, but already under decision-theoretic risk. First, decision makers tend to ignore low probabilities as irrelevant, focusing on the "more realistic", higher ones. This means that low, but significant probabilities for catastrophe are ignored, for example, so called "fat tails" in climate scenarios (Randall 2011, 77). Second, decision makers are often "myopic", placing higher weight on current costs than on future benefits, and avoiding high costs today. This often leads to even higher costs in the future. Third, disutilities might get calculated too optimistically, neglecting so-called "secondary effects" or "social amplifications", for example, the psychological and social effects of catastrophes (Sunstein 2007, 7). Lastly, since cost-benefit analysis provides such a clear view, there is a tendency to apply it even if the conditions for its application are not fulfilled. We tend to assume more than we know, and to decide according to the MEU criterion although no reliable probability information and/or no precise utility information is available. This so-called "tuxedo fallacy" is seen as a dangerous fallacy because it creates an "illusion of control" (Hansson 2008, 426–427).

Since PPs are seen as principles that address exactly such problems—drawing our attention to unlikely catastrophic possibilities, demanding action despite uncertainty, and that we consider the worst possible outcomes, and not assume more than we know—they gain indirect support from these arguments. ORM in its current form tempts us to apply it incorrectly and to neglect rational precautionary action. At least some sort of overarching PP that reminds us of correct practices seems necessary.

As a second point, it is argued that the regulatory practice of ORM has not only the "built-in" tendency to misapply its tools, but that it has fundamental flaws in itself which should be corrected by a PP. Randall (2011, 46–70) criticizes risk assessment in ORM on the grounds that it is typically built on simple models of the threatened system, for example, the climate system. Those neglect systemic risks like the possibility of feedback effects or sudden regime shifts. By depending on the law of large numbers, ORM is also not a decision framework that is suitable to deal with potential catastrophes, since they are singular events (Randall 2011, 52). Similarly, Chisholm and Clarke (1993, 112) argue that expected utility theory is only useful as long as "probabilities and possible outcomes are within the normal range of human experience". Examples of such probabilities and outcomes in the normal range of human experience are insurances like car and fire insurance: we have statistics about the probabilities of accidents or fires, and can calculate reasonable insurance premiums based on the law of large numbers. Furthermore, we have experience with how to handle them, and have institutions in place like fire departments. None of this is true for singular events like anthropogenic climate change. Consequently, it is argued that we cannot just leave ORM relatively unaltered, supporting it with a PP for decisions under uncertainty, and perhaps a more general, overarching PP as a normative guideline. Instead, it is asserted that we also have to reform the existing ORM framework in order to include precautionary elements.

Historical Arguments for Revising ORM

In the past, failures to take precautionary measures often resulted in substantial, widespread, and long-term harm to the environment and human health (Gee et al. 2013; Harremoës et al. 2001). This insight has been used to defend adopting a precautionary principle as a corrective to existing practices: for John (2007, 222), these past failures can be used as "inductive evidence" in an argument for reforming our regulatory policies. Whiteside (2006, 146) defends a PP as a product of social learning from past mistakes. According to Whiteside, these past mistakes reveal that (a) our knowledge about the influences of our actions on complex ecological systems is insufficient, and (b) that how decisions were reached was an important part of their inefficiency, leading to insufficient protection of the environment and human health. As such, to Whiteside, the PP generates a normative obligation to re-structure our decision-procedures (Whiteside 2006, 114). The most elaborate historical argument is made by Steel (2015, chapter 5). Steel's argument rests on the following premise:

> If a systematic pattern of serious errors of a specific type has occurred, then a corrective for that type of error should be sought. (Steel 2015, 91)

By critically examining not only cases of failed precautions and harmful outcomes, but also counter-examples of allegedly "excessive" precaution, Steel shows that such a pattern of serious errors in fact exists. Cases such as the ones described in "Late Lessons from Early Warnings" (Harremoës et al. 2001) demonstrate that continuous delays in response to emerging threats have frequently led to serious and persistent harms. Steel (2015, 74–77) goes on to examine cases that have been cited as examples of excessive precaution. He finds that in fact, often no regulation whatsoever was implemented in the first place. And in cases where regulations were put in place, they were mostly very restricted, had only minimal negative effects, and were relatively easily reversible. For example, one of the "excessive precautions" consisted in putting a warning label on products containing saccharine in the United States. According to Steel (2015, 82), the historical argument thus supports a PP as a corrective against a systematic bias that is entrenched in our practices. This bias emerges because there are informational and political asymmetries that make continual delays more likely than precautionary measures when there are trade-offs between short-term economic gain for an influential party against harms that are uncertain or distant in terms of space or time (or all three).

Epistemic Implications

The justifications presented so far all concern PPs aiming at the management of risks, that is, action-guiding interpretations. But we can also find discussions of a PP for the assessment of threats, so called "epistemic" PPs. It is not enough to just supply existing practices with a PP; clearly, risk assessment has to be changed, too, in order to be able to apply a PP. This means that uncertainties have to be taken seriously and to be communicated clearly, that we need to employ more adequate models which take into account the existence of systemic risks (Randall 2011, 77–78), that we need criteria to identify plausible (as opposed to "mere") possibilities, and so on. However, this is more a question of the implications of adopting a PP, not an expression of a genuine PP itself. Thus, these kinds of argument are either presuppositions for a PP, because we need to identify uncertain harms first in order to do something about them; or they are implications from a PP, because it is not admissible to conduct a risk assessment that makes it impossible to apply a PP.

Procedural Precaution

Authors who favor a procedural interpretation of PPs stress that they are concerned especially with decisions under conditions of uncertainty. They point out that while ORM, with its focus on cost-effectiveness and maximizing benefits, might be appropriate for conditions of decision-theoretic risk, the situation is fundamentally different if we have to make decisions under decision-theoretic uncertainty or even decision-theoretic ignorance. For example, Arcuri (2007, 20) points out that since PPs are principles particularly for decisions under decision-theoretic uncertainty, they cannot be prescriptive rules which tell us what the best course of action is—because the situation is essentially characterized by the fact that we are uncertain about the possible outcomes to which our actions can lead. Tickner (2001, 14) claims that this should lead to a redirection of the questions that are asked in environmental decision-making: the focus should be moved from the hazards associated with a narrow range of options, to solutions and opportunities. Thus, the assessment of alternatives is a central point of implementing PPs in procedural frameworks:

> In the end, acceptance of a risk must be a function not only of hazard and exposure but also of uncertainty, magnitude of potential impacts and the availability of alternatives or preventive options. (Tickner 2001, 122)

Although (economic) efficiency should not be completely dismissed and still should have its place in decision-making, proponents of a procedural PP proclaim that we should shift our aim in risk regulation from maximizing benefits to minimizing threats, especially in the environmental domain where harms are often irreversible (cf. Whiteside 2006, 75). They also advocate democratic participation, pointing out that a decision-making process under scientific uncertainty cannot be a purely scientific one (Whiteside 2006, 30–31; Arcuri 2007, 27). They thus see procedural interpretations of PPs as justified with respect to the goal of ensuring that decisions are made in a responsible and defensible way, which is especially important when there are substantial uncertainties about their outcomes.

Challenging the Underlying Value Assumptions

In addition to scientific uncertainty, Resnik (2003, 334) distinguishes another kind of uncertainty, which he calls “axiological uncertainty”. Both kinds make it difficult to implement ORM in making decisions. While scientific uncertainty arises due to our lack of empirical evidence, axiological uncertainty is concerned with our value assumptions. This kind of uncertainty can take on different forms: we can be unsure about how to measure utilities—in dollars lost/saved, lives lost/saved, species lost/saved, or something else? Then, we can be uncertain how to aggregate costs and benefits, and how to compare, for example, economic values with ecological ones. Values cannot always be measured on a common ordinal scale, much less on a common cardinal scale (as ORM requires, at least in some senses such as those including the use of a version of cost-benefit analysis). Thus, it is irrational to treat them as if they would fulfill this requirement (Aldred 2013; Thalos 2012). This challenges the value assumptions underlying ORM, and is seen as a problem that should be fixed by a PP.

Additionally, authors like Hansson (2005b, 10) object that it is essentially problematic that costs and benefits get aggregated without regard to who has them, and that person-related aspects like autonomy, or if a risk is willingly taken or imposed by others, are unjustly neglected.

To sum up, we can say that when the underlying value assumptions of ORM are challenged, the criticism pertains either to how values are estimated and assigned, or to the utilitarian decision criterion of maximizing overall expected utility is criticized. In both cases, we are arguably leaving the framework of rational choice and ORM, and moving toward genuine moral justifications for PPs.

4.3.2 Moral Justifications for Precaution

Some authors stress that, regardless of whether a PP is thought to supplement ordinary risk management (ORM) or whether it is a more substantive claim, a PP is essentially a moral principle, and has to be justified on explicitly moral grounds. (Note that depending on the moral position one holds, many of the considerations in 3.1 can also be seen as discussions of PPs from a moral standpoint; most prominently utilitarianism, since ORM uses the rule of maximizing expected utility.) They argue that taking precautionary measures under uncertainty is morally required, because otherwise we risk damages that are in some way morally unacceptable.

4.3.2.1 Environmental Ethics

PPs are often associated with environmental ethics, and the concept of sustainable development (Kaiser 1997; McKinney and Hill 2000; O'Riordan and Jordan 1995; Paterson 2007; Steele 2006; Westra 1997). Some authors take environmental preservation to be at the core of PPs. PP formulations as the Rio or the Wingspread PP emerged in a debate about the necessity to prevent environmental degradation, which explains why many PPs highlight environmental concerns. It seems plausible that a PP can be an important part of a broader approach to environmental preservation and sustainability (Ahteensuu 2008, 47). But it seems difficult to justify a PP with recourse to sustainability, since the concept itself is vague and contested. Indeed, when PPs have been discussed in the context of sustainability, they are often proposed as ways to operationalize the vague concept into a principle for policymaking, along with other principles like the "polluter pays" principle (Dommen 1993; O'Riordan and Jordan 1995). Thus, while PPs are partly motivated by the insight that our way of life is not sustainable, and that we should change how we approach environmental issues, it is difficult to justify them solely on such grounds. However, the hope is that a clarification of the normative (moral) underpinnings of PPs will help to justify a PP for sustainable development. In the following, we will see that it might make sense to take special precautions with respect to ecological issues, not only because they often are complex and might entail unresolvable uncertainties (Randall 2011, 64–70), but also because harm to the environment can affect many other moral concerns, for example, human rights and both international and intergenerational justice. As we will see, these moral issues might provide justifications for PPs on their own, without explicit reference to sustainability.

4.3.2.2 Harm-Based Justifications

PPs that apply to governmental regulatory decisions have been defended as an extension of the harm principle. There are different versions of the harm principle, but roughly it states that the government is justified in restricting citizens' individual liberty only to avoid harm to others. The application of the harm principle normally presupposes that certain conditions are fulfilled, for example, that the harms in question must be (1) involuntarily taken, (2) sufficiently severe and (3) probable, and (4) the prescribed measures must be proportional to the harms (cf. Jensen 2002; Petrenko and McArthur 2011). If these conditions are fulfilled, the prevention principle can be applied, prescribing proportional measures to prevent the harm in question from materializing. However, PPs apply to cases where we are unsure about the extent and/or the probability of a possible harm. Consequently, PPs are seen as a "clarifying amendment" (Jensen 2002, 44) which extends the normative foundation of the harm principle from prevention to precaution (Petrenko and McArthur 2011, 354): the impossibility of assigning probabilities does not negate the obligation to act as long as possible harms are severe enough and scientifically plausible. Even for the prevention principle, it holds that the more severe a threat is, the less probable it has to be in order to warrant preventive measures. Thus, it has been argued that the probability of high-magnitude harms becomes almost irrelevant, as long as they are scientifically plausible (Petrenko and McArthur 2011, 354–55). Additionally, some harm is seen as so serious that it warrants special precaution, for example, if it is irreversible or cannot be (fully) compensated (Jensen 2002, 49–50). In such situations, the government is justified in restricting liberties by, for example, prohibiting a technology, even if there remains uncertainty about whether or not the technology would actually have harmful effects.

A related idea is that governments have an institutional obligation not to harm the population, which overrides the weaker obligation to do good—meaning that it is worse if certain regulatory decisions of the government lead to harm than if they lead to foregone benefits (John 2007).

The question of what exactly makes a threat severe enough to justify the implementation of precautionary measures has also been discussed with reference to justice- and rights-based considerations.

4.3.2.3 Justice-Based Justifications

McKinnon (2009, 2012) presents two independent arguments for precautions, which both are justice-based. Those arguments are developed with respect to the possibility of a climate change catastrophe (CCC), and concern two alternative courses of action and their worst cases. The case of "Unnecessary Expenditure" means taking precautions which turn out to have been unnecessary, thereby wasting money which could have been spent for other, better purposes. "Methane Nightmare" describes the case of not taking precautions, leading to CCCs with catastrophic consequences, making survival on earth very difficult if not impossible. McKinnon argues that CCCs are uncertain in the sense that they are scientifically plausible, even though we cannot assign probabilities to them (McKinnon 2009, 189).

Playing it Safe

McKinnon's first argument for why uncertain yet plausible harm with the characteristics of CCCs justifies precautionary measures is called the "playing safe" argument. It is based on two Rawlsian commitments about justice (McKinnon 2012, 56): (1) That treating people as equals means (among other things) ensuring a distribution of (dis)advantage among them that makes the worst-off group as well off as possible, and (2) that justice is intergenerational in scope, governing relations across generations as well as within them.

McKinnon (2009, 191–92) argues that the distributive injustice would be so much higher if "Methane Nightmare" should materialize than if it came to "Unnecessary Expenditure" that we have to choose to take precautionary measures, even though we do not know how probable "Methane Nightmare" is. That is to say, such a situation warrants the application of the maximin-principle, because distributive justice in the sense of making the worst-off as well off as possible has lexical priority to maximizing the overall benefits for all. Choosing an option that has a far better best case, but, in the worst case, would lead to distributive injustice, over another option which might have a less-good best case, but where the worst case does not entail such distributive injustices, would be inadmissible.

Unbearable Strains of Commitment

As McKinnon notes, the "playing safe" justification only holds if one accepts a very specific understanding of distributive (in)justice. However, she claims to have an even more fundamental argument for precautionary measures in this context, which is also based on Rawlsian arguments concerning intergenerational justice, but does not rely on a specific conception of distributive justice. It is called the "unbearable strains of commitment" argument and is based on a combination of the "just savings" principle for intergenerational justice together with the "impartiality" principle. It states that we should not choose courses of actions that impose on future generations conditions which we ourselves could not agree to and which would undermine the bare possibility of justice itself (McKinnon 2012, 61). This justifies taking precautions against CCCs, since the worst case in that option is "Unnecessary Expenditure", which, in contrast to "Methane Nightmare" would not lead to justice-jeopardizing consequences.

4.3.2.4 Rights-Based Justifications

Strict precautionary measures concerning climate change have been demanded based on the possible rights violations that such climate change might entail. For example, Caney (2009) claims that although other benefits and costs might be discounted, human rights are so fundamental that they must not be discounted. He argues that the possible harms involved in climate change justify precautions: unmitigated climate change entails possible outcomes which would lead to serious or catastrophic rights violations, while a policy of strict mitigation would not involve a loss of human rights—at least not if it is carried out by the affluent members of the world. Additionally, "business as usual" from the affluent would mean to gamble with the conditions of those who already lack fundamental rights protection, because the negative effects of climate change would come to bear especially in poor countries. Moreover, the benefits of taking the risk of catastrophic climate change outcomes would almost entirely result for the risk-takers, not the risk-bearers (Caney 2009, 177–79). If we extrapolate from this concrete application, the basic justification for precaution seems to be: if a rights violation is plausibly possible, and there are ways to avoid this possibility by choosing another course of action which does not involve the plausible possibility of rights violations, then we have to choose the second option. It does not matter how likely it is that the rights violations shall happen; as long as they are plausible, we have to treat them as if they would materialize with certainty.

Thus, in this interpretation, precaution means making sure that no rights violations happen, even if we (because of uncertainty) "run the risk" of doing more than what would have been necessary—as long as we don't have to jeopardize our own rights in order to do so.

4.3.2.5 Ethics of Risk and Risk Impositions

Some authors see the PP as an expression of a problem with what they call standard ethics (Hayenhjelm and Wolff 2012, e28). According to them, standard ethical theories with their focus on evaluations of actions and their outcomes under conditions of certainty fail to keep up with the challenges posed by technological development. PPs are then placed in the broader context of developing and defending an ethics of risk, i.e., a moral theory about the permissibility of risk impositions. Surprisingly, so far there are few explicit connections between the discussion of the ethics of risk impositions (see for example Hansson 2013; Lenman 2008; Suikkanen 2019) and the discussion of PPs.

One exception is Munthe (2011), who argues that before we can formulate an acceptable and intelligible PP, we first need at least the basic structure of an ethical theory that deals directly with issues of creating and avoiding risks of harm. In Chap. 5 of his book, Munthe sets out to develop such a theory, which focuses on the responsibility of a decision, specifically, responsibility as a property of decisions: decisions and risk impositions may be morally appraised in their own right. When one does not know what the outcome of a decision will be, it is important to make responsible decisions, i.e., decisions that can still be defended as having been responsible given the information one had at the time the decision was made, even if the outcome is wrong. However, even though Munthe's discussion starts out from the PP, he ultimately concludes that we do not need a PP, but rather a policy that expresses a proper degree of precaution:

> What is needed is plausible theoretical considerations that may guide decision makers also employing their own judgement in specific cases. We do not need a precautionary principle, we need a policy that expresses a proper degree of precaution. (Munthe 2011, 164)

Thus, the idea seems to be that while a fully developed ethics of risk will justify demands commonly associated with PPs, it ultimately will replace the need for a PP.

4.4 Main Objections and Possible Rejoinders

This section presents the most frequent and the most important objections and challenges PPs face. They can be roughly divided into three groups. The first argues that there are fundamental conceptual problems with PPs, which make them unable to guide our decisions. The second claims that PPs, in any reasonable interpretation, are superfluous and can be reduced to existing practices done right. The third rejects PPs as irrational, saying that they are based on unfounded fears and that they contradict science, leading to undesirable consequences. While some objections are aimed at specific PP proposals, others are intended as arguments against PPs in general. However, even the latter typically hold only for specific interpretations. This section briefly presents the main points of these criticisms, and then discusses how they might be answered.

4.4.1 PPs Cannot Guide our Decisions

There are two main reasons why PPs are seen as unable to guide us in our decision-making: they are rejected either as incoherent, or as being vacuous and devoid of normative content.

Objection: PPs are Incoherent

One frequent criticism, most prominently advanced by Sunstein (2005b), is that a "strong PP" leads to contradictory recommendations and would therefore be paralyzing for our decision-making. He understands "strong PP" as a very demanding principle which states that "regulation is required whenever there is a possible risk to health, safety, or the environment, even if the supporting evidence remains speculative and the economic costs of regulation are high" (Sunstein 2005b, 24). The problem is that every action poses such a possible risk, and thus both regulation and non-regulation would be prohibited by the "strong PP", resulting in paralysis (Sunstein 2005b, 31). Hence, "strong PP" is rejected as an incoherent decision rule, because it leads to contradictory recommendations.

Peterson (2006) makes another argument that rejects PPs as incoherent. He claims that he can prove formally as well as informally that every serious PP formulation is logically inconsistent with reasonable conditions of rational choice, and should therefore be given up as a decision rule (Peterson 2006, 597).

Rejoinder

Both criticisms have been rejected as being based on a skewed interpretation of the PP. In the case of Sunstein's argument, he is attacking a straw-man. His critique of the "strong PP" as paralyzing relies on two assumptions which are not made explicit, namely (a) that a PP is invoked by any and all risks, and (b) that risks of action and inaction are typically equally balanced (Randall 2011, 20). However, this is an atypical PP interpretation. Most formulations make explicit reference to severe dangers, meaning that not just any possible harm, no matter how small, will invoke a PP. And, as the case studies in Harremoës et al. (2001) illustrate, the possible harms from action and inaction—or, more precisely, regulation or no regulation—are typically not equally balanced (see also Steel 2015, chapter 9). Still, Sunstein's critique calls attention to the important point of risk-risk trade-offs, which every sound interpretation and application of a PP has to take into account: taking precautions against a possible harm should not lead to an overall higher level of threat (Randall 2011, 84–85). Nevertheless, there seems to be no reason why a PP should not be able to take this into account, and the argument thus fails as a general rejection of PPs.

Similarly, it can be contested whether Peterson's PP formalization is a plausible PP candidate: he presupposes that we can completely enumerate the list of possible outcomes, that we have rational preferences that allow for a complete ordering of the outcomes, and that we can estimate at least the relative likelihood of the outcomes. As Randall (2011, 86) points out, this is an ideal setup for ordinary risk management (ORM), and the three conditions for rational choice that Peterson cites, and with which he shows his PP to be inconsistent, have their place in the ORM-framework. Thus, one can object that it is not very surprising if a PP, which is designed especially for situations in which ideal conditions are *not* met, does not do very well under ideal conditions.

Objection: PPs are Vacuous

On the other hand, it is argued that if a PP is attenuated in order not to be paralyzing, it becomes such a weak claim that it is essentially vacuous. Sunstein (2005b, 18) claims that weaker formulations of PPs are, although not incoherent, trivial: they merely state that lack of absolute scientific proof is no reason for inaction, which, according to Sunstein, has no normative force because everyone is already complying with it. Similarly, McKinnon (2009) takes a weak PP formulation to state that precautionary measures are permissible, which she also rejects as a hollow claim, stating that everyone could comply with it without ever taking any precautionary action.

Additionally, PPs are rejected as vacuous because of the multitude of formulations and interpretations. Turner and Hartzell (2004), examining different formulations of PPs, come to the conclusion that they are all beset with unclarity and ambiguities. They argue that there is no common core to the different interpretations, and that the plausibility of a PP actually rests on its vagueness. This makes it unsuitable as a guide for decision-making. Similarly, Peterson (2007a, 306) states that such a "weak" PP has no normative content and no implications for what ought to be done. He claims that in order to have normative content, a PP would need to give us a precise instruction what to do for each input of information. By formulating a minimal normative PP interpretation and showing that it is incoherent, he argues that there cannot be a PP with normative content.

Rejoinder

Firstly, let us address the criticism that PPs are vacuous because they express a claim that is too weak to have any impact on decision-making. Against this, Steel (2013, 2015) has argued that even if these supposedly "weak" or "argumentative" principles do not directly recommend a specific decision, they nonetheless have an impact on the decision-making process if taken seriously. He interprets them as a meta-principle that puts constraints on what decision rules should be used, namely, none that would lead to inaction in the face of uncertainty. Since, for example, cost-benefit analysis needs numerical probabilities to be applicable, the Meta PP will recommend against it in situations where no such probability information is available. This is a substantial constraint, meaning that the Meta PP is not vacuous. One can reasonably doubt that Sunstein is right that everyone follows such an allegedly "weak" principle anyway. There are many historical cases where there was some positive evidence that an activity caused harm, but the fact that the activity–harm link had not been irrefutably proven was used to argue against regulatory action (Gee et al. 2013; Harremoës et al. 2001). Thus, in cases where no proof, or at least no reliable probability information, concerning the possibility of harm is available, uncertainty is often used as a reason to not to take precautionary action. Additionally, this criticism clearly does not concern all forms of PPs, and only amounts to a full-fledged rejection of PPs if combined with the claim that so-called "stronger" PPs. which are not trivial, will always be incoherent. And both Sunstein (2005b) and McKinnon (2009, 2012) do propose other PPs which express a stronger claim, albeit with a restricted scope (for example, only pertaining to catastrophic harm, or damage which entails specific kinds of injustice). This form of the "vacuous" objection can thus be seen not as an attack on the general idea of PPs, but more as the demand that the normative obligation they express should be made clear in order to avoid downplaying it.

Let us now consider the other form of the objection, namely the claim that PPs are essentially vague and that there cannot be a precise formulation of a PP that is both action-guiding and plausible. It is true that, so far, there does not seem to exist a "one size fits all" PP that yields clear instructions for every input and that captures all the ideas commonly associated with PPs. However, even if this were a correct interpretation of what a "principle" is (which many authors deny, compare for example Randall 2011, 97), it is not the only one. Peterson (2007a) presumes that only a strict "if this, then that" rule can have normative force, and consequently be action-guiding. In contrast, other authors stress the difference between a principle and a rule (Arcuri 2007; Fisher 2002; Randall 2011). According to them, while rules specify precise consequences that follow automatically when certain conditions are met, principles express normative obligations that need to be specified according to different contexts, and that need to be implemented and operationalized in rules, laws, policies, and so on (Randall 2011, 97). When authors are rejecting PPs as incoherent (see the previous paragraph), they might sometimes make the same mistake, confusing a general principle that needs to be specified on a case-by-case basis with a stand-alone decision rule that should fit for any and all cases.

As for PPs being essentially vague: this criticism seems to presuppose that in order to formulate a clarified PP, we have to capture and unify everything that is associated with it. However, explicating a concept in a way that clarifies it and captures as many of the ideas associated with it as possible does not mean that we have to preserve *all* of the ideas commonly associated with it. The same is true for explicating a principle such as a PP. Additionally, this article shows that many different ways of interpreting PPs in a precise way are possible, and not all of them exclude each other.

4.4.2 PPs are Redundant

Some authors reject PPs by arguing that they are just a narrow and complicated way of expressing what is already incorporated into established, more comprehensive approaches. For example, Bognar (2011) compares Gardiner's (2006) "Rawlsian Core PP" interpretation with what he calls a "utilitarian principle" which consists of a combination of the principles of indifference and that of maximizing expected utility. He concludes that this "utilitarian principle" does lead to the same results as the RCPP in the cases where the RCPP applies, but, contrary to it, this "utilitarian principle" is not restricted to such a narrow range of cases. His conclusion is that we can dispose of PPs, at least in formulations of maximin (Bognar 2011, 345).

In the same vein, (Peterson 2007a, 600) asserts that if formulated in a consistent way, a PP would not be different from the "old" rules for risk-averse decision-making, while other authors have shown that we can use existing ordinary risk-management (ORM) tools to implement a PP (Farrow 2004; Gollier et al. 2001). This allegedly would make PPs redundant (Randall 2011, 25; 87).

Rejoinder
Particularly against the criticism from Bognar (2011), one can counter that his "utilitarian principle" falls victim to the so-called "tuxedo fallacy" (Hansson 2008). Using the principle of indifference, that is, treating all outcomes as equally probable when one does not have enough information to assign reliable probabilities, can be seen as creating an "illusion of control" by assuming that as long as no probability information is available, all outcomes are equally probable. It neither pays sufficient attention to catastrophic harms, nor takes the special challenges of decision-theoretic uncertainty adequately into account.

More generally, one can make the following point: even though there might be plausible ways in which we can translate a PP into the ORM-framework and implement it using ORM-tools, there is more to it than that. Even if we use ORM-methods to implement precaution, in the end this might still be based on a normative obligation to enact precautionary measures. This obligation has to be spelled out, because ORM *can allow* for precaution, but does not demand it in itself (and, as a regulatory practice, tends to neglect it).

4.4.3 PPs are Irrational

The last line of criticism accuses PPs of being based on unfounded fears, or expressing cognitive biases, and therefore leading to decisions with undesirable and overall harmful consequences.

Objection: Unfounded Panic
One criticism that is especially frequent in discussions aimed at a broader audience is that PPs give way to unrestrained regulation, because they can be invoked by uncertain harm. Thereby, the argument goes, PPs pose a danger of unnecessary expenditures to reduce insignificant risks, or of foregone benefits by regulating or prohibiting potentially beneficial activities, and are prone to being exploited, for example by interest groups or for protectionism in international trade (Peterson 2006). A PP would stifle innovation, resulting in an overall less safe society: many (risk-reducing) beneficial innovations of the past were only possible because risks have been taken (Zander 2010, 9) and technical innovation takes place in a process of trial-and-error, which would be seriously disturbed by a PP (Graham 2004, 5).

Such critics see these as possible consequences of PPs because PPs do not require scientific certainty in order to take action, and they interpret this as making merely speculative harm a reason for strict regulation. Thus, science would be marginalized or even rejected as a basis for decision-making, giving way to the cognitive biases of ordinary people.

Objection: Cognitive Biases
Sunstein (2005b, chapter 4) claims that PPs are based on the cognitive biases of ordinary people, which tend to systematically mis-asses risks. By reducing the importance of scientific risk-assessment and marginalizing the role of experts, decisions resulting from the application of a PP will be influenced by these biases and result in negative consequences, the criticism goes.

Rejoinder
As has been pointed out by Randall (2011, 89), these criticisms seem to be misguided. Lower standards of evidence do not mean no standards at all. It is surely an important challenge for the implementation of a PP to find a way to define plausible possibilities, but this requires by no means less science. Instead, as Sandin et al. (2004) point out, more, and different scientific approaches are needed. Uncertainties need to be communicated more clearly and tools need to be developed that allow taking uncertainties better into account. For decisions where we lack scientific information, but great harms are possible, ways need to be found in which public concerns can be taken into consideration (Arcuri 2007, 35). This, however, seems more a question of implementation than of the formulation or the justification of a PP.

4.5 Recapitulation

Section 4.4 shows that there are compelling rejoinders to the general criticisms of PPs. The question is then whether there is already a specific candidate that is able to answer these criticisms while meeting the following minimal requirements for a successful clarification and formulation of a PP, which I identified in the beginning:

- The conditions under which the PP applies and the requirements that follow from it have to be clear, and
- since it expresses a normative claim, it has to be stated on which grounds the principle is justified.

As this survey shows, there are very different ways to formulate and defend a PP. Is there already a proposal for a PP that meets these two minimal requirements?

On the basis of the present survey, the most promising candidates are the Rawlsian Core Precautionary Principle of Gardiner (2006), the integrated risk-regulation framework of Randall (2011), the tripartite proposal of Steel (2015), and the Catastrophic Harm Precautionary Principle of Hartzell-Nichols (2017). They all give relatively clear conditions under which the PP proposal in question is supposed to apply, and specify what would follow from it, or, respectively, how to determine it.

In part, how action-guiding these proposals can be will also depend on how they are implemented. Here, epistemic and procedural questions become important. However, I argue that the latter are not themselves specific PP interpretations, but rather implications of a PP, or respectively presuppositions for the implementation of a PP. Thus, which PP is ultimately adopted will also have implications for how our epistemic and procedural practices should be reformed. The question then is which PP we actually should adopt.

This brings us to the second requirement, the question of justification. Each of the selected candidates is supported by its authors, e.g., based on arguments that show that the verdicts of the principle fit with judgments that we hold, by answering criticisms, by developing the proposal against the weaknesses of another approach, or by arguing that there is an uncontroversial moral duty to avoid catastrophes.[3] Perhaps the most comprehensive arguments are those made by Randall (2011), who develops his PP as a result of an in-depth engagement with current risk-regulation practices, and Steel (2015), who develops his PP through a detailed process of either accounting for or refuting claims about PPs, uses case studies to show the plausibility of the implications of his proposal, and brings forward an historical argument that supports the demand for a PP independently of controversial presumptions about ethical theories.

This leaves us with two questions: firstly, how can we comparatively assess these proposals and decide which one we actually should adopt? Not surprisingly in the context of this book, I suggest that reflective equilibrium is a suitable method for this task. The case study of justifying a PP can also be seen as a way of spelling out how different candidates would have to be compared.

Secondly, is there a way to develop a PP as a substantial moral principle? A lot of work already has been done on interpretations of PPs as principles of rational choice that are intended for public policy-making. In this context, it can be desirable to have a principle that does not rely too heavily on moral commitments. However, if it were possible to develop a defensible principle that tells us why taking precautions is not only rationally or prudentially required, but why there is a *moral* obligation to take precautions, this would give additional urgency to the demand for precautions. This is especially true for intergenerational cases like climate change.

Thus, I argue that while there are already answers that come to close to a satisfying proposal for a PP, the urgent moral questions of cases like climate change make it especially worthwhile to reconsider the formulation and justification of a PP that is suitable also for intergenerational cases.

---

[3] "Uncontroversial" meaning that catastrophe is defined in a way such that every moral position will recommend taking action to avoid it. Compare Hartzell-Nichols (2012, 161): "If we owe our future selves or future people anything, it seems plausible that we have a prima facie obligation to take precautionary measures against foreseeable catastrophes. [...] While we cannot take precautionary measures against every possible threat of harm, there is something to this intuition. We at least ought to take precautionary measures against the very worst kind of outcomes, namely those that would be catastrophic."

In the next chapter, I describe the design of a case study for the method of reflective equilibrium (RE). This application of RE has the pragmatic-epistemic objective of justifying an action-guiding moral principle that is applicable to the subject matter of precaution and precautionary decision-making. The process of adjusting commitments and system in order to make progress towards this goal is described in Chaps. 6–8. As the case study will show, a rights-based precautionary principle seems like a promising candidate for a PP that takes substantial moral claims into account.

Acknowledgments This chapter is based on an encyclopedia article that is published online in the Internet Encyclopedia of Philosophy: https://www.iep.utm.edu/pre-caut/.

References

Ahteensuu M (2008) In Dubio Pro Natura? A philosophical analysis of the precautionary principle in environmental and health risk governance. PhD thesis, University of Turku, Turku

Aldred J (2013) Justifying precautionary policies: Incommensurability and uncertainty. Ecol Econ 96:132–140. https://doi.org/10.1016/j.ecolecon.2013.10.006

Arcuri A (2007) The case for a procedural version of the precautionary principle erring on the side of environmental preservation. SSRN Scholarly Paper ID 967779. Social Science Research Network, Rochester

Arrow KJ, Fisher AC (1974) Environmental preservation, uncertainty, and irreversibility. Q J Econ 88(2):312–319. https://doi.org/10.2307/1883074

Bognar G (2011) Can the maximin principle serve as a basis for climate change policy? Monist 94(3):329–348. https://doi.org/10.5840/monist201194317

Buchak L (2013) Risk and Rationality. OUP Oxford, Oxford

Caney S (2009) Climate change and the future: discounting for time, wealth, and risk. J Soc Philos 40(2):163–186

Chisholm AH, Clarke HR (1993) Natural resource management and the precautionary principle. In: Dommen E (ed) Fair principles for sustainable development: essays on environmental policy and developing countries, pp 109–122

Dommen E (ed) (1993) Fair principles for sustainable development: essays on environmental policy and developing countries. Edward Elgar, London

Farrow S (2004) Using risk assessment, benefit-cost analysis, and real options to implement a precautionary principle. Risk Anal 24(3):727–735

Fisher E (2002) Precaution, precaution everywhere: developing a common understanding of the precautionary principle in the European community. Maastricht Journal of European and Comparative Law 9(1):7–28

Gardiner SM (2006) A core precautionary principle. J Polit Philos 14(1):33–60

Gee D, Grandjean P, Hansen SF, van denHove S, MacGarvin M, Martin J, Nielsen G, Quist D, Stanners D (2013) Late lessons from early warnings: science, precaution, innovation. Technical report. European Environment Agency, London

Gollier C, Moldovanu B, Ellingsen T (2001) Should we beware of the precautionary principle? Econ Policy 16(33):303–327

Graham JD (2004) The Perils of the precautionary principle: lessons from the American and European experience, vol 818. Heritage Foundation, Washington

Hansson SO (1997) The limits of precaution. Found Sci 2(2):293–306

Hansson SO (2005a) Decision theory: a brief introduction

Hansson SO (2005b) Seven myths of risk. Risk Manage 7(2):7–17

Hansson SO (2008) From the casino to the jungle. Synthese 168(3):423–432. https://doi.org/10.1007/s11229-008-9444-1
Hansson SO (2013) The ethics of risk: ethical analysis in an uncertain world. Palgrave Macmillan, New York
Harremoës P, Gee D, MacGarvin M, Stirling A, Keys J, Wynne B, Vaz SG (eds) (2001) Late lessons from early warnings: the precautionary principle 1896–2000. Office for Official Publications of the European Communities, Luxembourg
Harris J, Holm S (2002) Extending human lifespan and the precautionary paradox. J Med Philos 27(3):355–368
Harsanyi JC (1975) Can the maximin principle serve as a basis for morality? A Critique of John Rawls's Theory. Am Polit Sci Rev 69(2):594–606. https://doi.org/10.2307/1959090
Hartzell-Nichols L (2012) Precaution and solar radiation management. Ethics, Policy and Environment 15(2):158–171. https://doi.org/10.1080/21550085.2012.685561
Hartzell-Nichols L (2013) From 'The' Precautionary Principle to Precautionary Principles. Ethics, Policy and Environment 16(3):308–320
Hartzell-Nichols L (2017) A climate of risk: precautionary principles, catastrophes, and climate change. Routledge, New York
Hayenhjelm M, Wolff J (2012) The moral problem of risk impositions: a survey of the literature. Eur. J. Philos. 20(S1):E26–E51
Iverson T, Perrings C (2012) Precaution and proportionality in the management of global environmental change. Glob Environ Chang 22(1):161–177. https://doi.org/10.1016/j.gloenvcha.2011.09.009
Jensen KK (2002) The Moral Foundation of the Precautionary Principle. J Agric Environ Ethics 15(1):39–55. https://doi.org/10.1023/A:1013818230213
John SD (2007) How to take deontological concerns seriously in risk—cost—benefit analysis: a re-interpretation of the precautionary principle. J Med Ethics 33(4):221–224
John S (2010) In defence of bad science and irrational policies: an alternative account of the precautionary principle. Ethical Theory Moral Pract 13(1):3–18
Jonas H (1979) Das Prinzip Verantwortung: Versuch einer Ethik für die technologische Zivilisation, 1st edn. Suhrkamp Verlag, Frankfurt am Main
Kaiser M (1997) Fish-farming and the precautionary principle: context and values in environmental science for policy. Found Sci 2(2):307–341
Lemons J, Shrader-Frechette K, Cranor C (1997) The precautionary principle: scientific uncertainty and type I and type II errors. Found Sci 2(2):207–236
Lenman J (2008) Contractualism and risk imposition. Politics, Philosophy and Economics 7(1):99–122. https://doi.org/10/fqkwg3
Manson NA (2002) Formulating the precautionary principle. Environ Ethics 24(3):263–274
McKinney WJ, Hill HH (2000) Of sustainability and precaution: the logical, epistemological, and moral problems of the precautionary principle and their implications for sustainable development. Ethics and the Environment 5(1):77–87
McKinnon C (2009) Runaway climate change: a justice-based case for precautions. J Soc Philos 40(2):187–203
McKinnon C (2012) Climate change and future justice: precaution, compensation and triage. Routledge, London
Munthe C (2011) The price of precaution and the ethics of risk, the international library of ethics, law and technology, vol 6. Springer, Berlin
O'Riordan T, Jordan A (1995) The precautionary principle in contemporary environmental politics. Environmental Values 4(3):191–212
Osimani B (2013) An epistemic analysis of the precautionary principle. Dilemata: International Journal of Applied Ethics 5(11):149–167
Paterson J (2007) Sustainable development, sustainable decisions and the precautionary principle. Nat Hazards 42(3):515–528. https://doi.org/10.1007/s11069-006-9071-4

Peterson M (2003) Transformative decision rules. Erkenntnis 58(1):71–85
Peterson M (2006) The precautionary principle is incoherent. Risk Anal 26(3):595–601
Peterson M (2007a) The precautionary principle should not be used as a basis for decision-making. EMBO Rep 8(4):305–308. https://doi.org/10.1038/sj.embor.7400947
Peterson M (2007b) Should the precautionary principle guide our actions or our beliefs? J Med Ethics 33(1):5–10. https://doi.org/10.1136/jme.2005.015495
Petrenko A, McArthur D (2011) High-stakes gambling with unknown outcomes: justifying the precautionary principle. J Soc Philos 42(4):346–362
Randall A (2011) Risk and Precaution. Cambridge University Press, New York
Rawls J (2001) Justice as fairness: a restatement. Harvard University Press, Belknap
Resnik DB (2003) Is the precautionary principle unscientific? Studies in History and Philosophy of Science Part C: Studies in History and Philosophy of Biological and Biomedical Sciences 34(2):329–344
Resnik DB (2004) The precautionary principle and medical decision making. J Med Philos 29(3):281–299
Sandin P (1999) Dimensions of the precautionary principle. Hum Ecol Risk Assess Int J 5(5):889–907
Sandin P (2004) Better safe than sorry: applying philosophical methods to the debate on risk and the precautionary principle. PhD thesis, Stockholm
Sandin P (2007) Common-sense precaution and varieties of the precautionary principle. In: Lewens T (ed) Risk: philosophical perspectives, London and New York, pp 99–112
Sandin P (2009) A new virtue-based understanding of the precautionary principle. In: Ethics of Protocells: Moral and Social Implications of Creating Life in the Laboratory, pp 88–104
Sandin P, Hansson SO (2002) The default value approach to the precautionary principle. Hum Ecol Risk Assess Int J 8(3):463–471. https://doi.org/10.1080/10807030290879772
Sandin P, Peterson M, Hansson SO, Rudén C, Juthe A (2002) Five charges against the precautionary principle. J Risk Res 5(4):287–299
Sandin P, Bengtsson BE, Bergman Å, Brandt I, Dencker L, Eriksson P, Förlin L, Larsson P, Oskarsson A, Rudén C, Södergren A, Woin P, Hansson SO (2004) Precautionary defaults—a new strategy for chemical risk management. Hum Ecol Risk Assess 10(1):1–18
Science and Environmental Health Network (SEHN) (1998) The Wingspread Consensus Statement on the Precautionary Principle. https://www.sehn.org/sehn/wingspread-conference-on-the-precautionary-principle
Steel D (2011) Extrapolation, uncertainty factors, and the precautionary principle. Studies in History and Philosophy of Science Part C: Studies in History and Philosophy of Biological and Biomedical Sciences 42(3):356–364
Steel D (2013) The precautionary principle and the dilemma objection. Ethics, Policy and Environment 16(3):321–340
Steel D (2015) Philosophy and the Precautionary Principle. Cambridge University Press, Cambridge
Steele K (2006) The precautionary principle: A new approach to public decision-making? Law, Probability and Risk 5(1):19–31. https://doi.org/10.1093/lpr/mgl010
Suikkanen J (2019) Ex Ante and Ex Post Contractualism: A Synthesis. J Ethics 23(1):77–98. https://doi.org/10/ggjn22
Sunstein CR (2005a) Irreversible and Catastrophic. Cornell Law Rev 91:841–897
Sunstein CR (2005b) Laws of fear: beyond the precautionary principle. Cambridge University Press, New York
Sunstein CR (2007) The catastrophic harm precautionary principle. Issues in Legal Scholarship 6(3):1–29
Sunstein CR (2009) Worst-Case Scenarios. Harvard University Press, Harvard
Thalos M (2012) Precaution has its reasons. In: Kabasenche W, O'Rourke M, Slater M (eds) Topics in contemporary philosophy 9: the environment, philosophy, science and ethics. MIT Press, Cambridge, pp 171–184

Tickner JA (2001) Precautionary assessment: a framework for integrating science, uncertainty, and preventive public policy. In: Freytag E, Jakl T, Loibl G, Wittmann M (eds) The role of precaution in chemicals policy, diplomatische akademie wien, pp 113–127
Turner D, Hartzell L (2004) The lack of clarity in the precautionary principle. Environmental Values 13(4):449–460
United Nations Conference on Environment and Development (1992) Rio declaration on environment and development. https://undocs.org/en/A/CONF.151/26/Rev.1(vol.I)
Westra L (1997) Post-normal science, the precautionary principle and the ethics of integrity. Found Sci 2(2):237–262
Whiteside KH (2006) Precautionary politics: principle and practice in confronting environmental risk. MIT Press, Cambridge
Zander J (2010) The application of the precautionary principle in practice: comparative dimensions. Cambridge University Press, Cambridge

Chapter 5
Justifying a Precautionary Principle with Reflective Equilibrium: Design of a Case Study

Is reflective equilibrium (RE) a method that can be used in an insightful and fruitful way to justify principles or theories? In order to answer this question, and to gain further insights into the applicability of RE, I will conduct a case study in which I test whether RE can be used to formulate and justify a precautionary principle (PP). The present chapter describes the setup of this case study, whereas the application itself takes place in Chaps. 6–8.

5.1 Objectives and Overview

In Chap. 3, I proposed to spell out the method of reflective equilibrium as starting from an initial position and then proceeding in two alternating steps of adjusting commitments and system. In order to apply RE, one has to identify the elements of the initial position, and to specify the criteria of reflective equilibrium with respect to the particular justificatory project.

The first step is thus to clarify my pragmatic-epistemic objective and the subject matter, before specifying the method and describing the input. As this is a *case study* for reflective equilibrium as a method, is has two goals: one that is pursued within the application of RE, and one that is pursued with it, as a case study for RE.

Pragmatic-Epistemic Objective in the Case Study: Justifying an action-guiding moral principle that is applicable to the subject matter of precaution and precautionary decision-making (see Chap. 4).

Objective of the Case Study: Testing whether, and how, RE can be implemented as a method; and what we can learn about the theoretical foundations of RE by putting it into practice.

Having these two goals has certain consequences for the case study, as the method and its applicability (goal 2) is the main concern. To make the application of RE feasible and comprehensible, I will work with plausible simplifications and

T. Rechnitzer, *Applying Reflective Equilibrium*, Logic, Argumentation & Reasoning 27, https://doi.org/10.1007/978-3-031-04333-8_5

stipulations with respect to the content of the case study, e.g., only taking into account a very limited amount of (empirical) background information and only examining a restricted (but hopefully exemplary) set of commitments. Because of these restrictions, we cannot expect that a justified precautionary principle will follow. But the general structure and process should be exemplary and help to identify needs for modification, i.e., how to continue to work towards a justified position.

The structure of this chapter is as follows: I start by specifying the method of RE in Sect. 5.2, i.e., by concretizing the two RE steps through defining measures for the RE criteria. I then describe the selection of initial input commitments in Sect. 5.3, elements of the background in Sect. 5.4, my preliminary selection of theoretical virtues in Sect. 5.5, and candidates for the system in Sect. 5.6. Section 5.7 recapitulates the main points of the setup, and sketches the way ahead.

The elements of an RE process can quickly become hard to keep in mind. To keep the description manageable and comprehensible, I only describe exemplary or relevant aspects of the setup. The complete list of all the elements of the setup and of the RE process can be found in the appendix starting on p. 245.

5.2 Specifying the Criteria and Steps of Reflective Equilibrium

In Chap. 3, I developed a methodology for obtaining a method of RE based on the theoretical conception described in Chap. 2. I suggested that the RE process of adjustments can be structured in the form of two alternating kinds of steps:

Adjusting Commitments Keeping the system constant, find the set of commitments that maximizes the combination of (i) agreement with the current system, (ii) independent credibility, (iii) respect for input commitments, and (iv) support from background theories.

Adjusting the System Keeping the current set of commitments constant, find a system that maximizes the combination of (i) agreement with the current system, (ii) theoretical virtues, and (iii) support from background theories.

To specify the method for particular pragmatic-epistemic projects, we thus have to specify the various RE criteria, i.e., to define how they should be measured and how potential trade-offs should be handled. The following three main tasks were identified in Chap. 3:

- Define the RE criteria as exactly as possible while keeping them informative enough for the project at hand (i.e., the subject matter and pragmatic-epistemic objective in question, as well as the available resources);
- Give a preliminary weighting of the different criteria, noting if any of them are more important with respect to the pragmatic-epistemic objective;

- Concretize the two alternating steps of the process by inserting the so-defined criteria.

When specifying the criteria, my focus is on obtaining implementable criteria which are assessable in the practical application of the case study without making the process too technical. The goal is to work with plausible approximations of the RE criteria which can, based on the results of the case study, also serve as the basis for further elaboration and more refined specifications.

I will bracket the assessment of support from background theories at each step. Instead, I will use background theories as potential tie-breakers in case of trade-offs that are difficult to resolve, and to assess relatively well-advanced positions with respect to whether or not they are in a state of reflective equilibrium.

The other criteria I specify as follows:

Independent Credibility of Commitments Each commitment is assigned a weight that gives a rough indication of its (independent) credibility: commitments either have a low, medium, or high weight. This ranking is only ordinal, i.e., it expresses neither that two commitments with a high weight necessarily have the exact same degree of independent credibility, nor that the difference between low and medium is the same as the difference between medium and high.

Agreement between System and Commitments I specify the relation of agreement as **Account**, which is measured between a candidate system S_n and the set of current (explicit) commitments C_n. To measure account, a value is assigned for each commitment $c \in C_n$, depending on the kind of relation between S_n and c:

conflict (S_n implies $\neg c$): -2
consistent non-account (neither c nor $\neg c$ is implied by S_n): -0.5
partial account (S_n implies part of c): $+1$
full account (S_n implies c): $+2$
A weighted sum is then formed by first multiplying each such value with another value depending on the weight assigned to c:
low weight: $*1$
medium weight: $*2$
high weight: $*3$

A commitment is fully accounted for by the system if it can be inferred from the system via deductive or non-deductive valid arguments. These arguments can include background information or be supported by background theories. A commitment can be partially accounted for if a part of it can be inferred from the system. For example, this is possible in the case of general commitments: in order to fully account for a general commitment, the system would have to allow us to infer everything that can also be inferred from the general commitment. For example, if you have the commitment "You should never lie", then the principle "One should not lie if it will harm another person" will partially account for the commitment. However, to count as partial account, it is important that the principle stays silent on the part of the commitment that it does not account for: if the principle were "One should not lie if *and only if* it will harm another person", it would conflict with the commitment that you should never lie.

I assigned the numerical values based on the following considerations: we are aiming for full agreement between system and commitments, meaning that full account should be valued highest and conflict should receive the highest penalty. While consistency is a necessary condition for agreement, I decided to assign a small penalty for consistent non-account, as we want there to be some positive connection between system and commitments, i.e., something more than mere consistency. (Consistent) partial account is valuable—it can be an indicator that the current system gets *something* right, but of course it is less valuable than full account.

While the weights of the commitments are ordinal, as stated above, I still decided to assign numerical values to them in order to be able to take them into consideration when measuring account. Thus, they are effectively measured on an interval scale, but one has to take this measurement with a grain of salt—this account function only works as a rough indicator of how well competing candidates for the system are able to account for a given set of current commitments.

Respecting Input Commitments When adjusting a set of current commitments C_n with respect to a current system S_n, then each adjustment towards increasing account is *lexically constrained* by the criterion that current commitments have to respect input commitments: an input commitment can only be adjusted if it can be plausibly argued that its independent credibility is outweighed or negated by other considerations.[1]

An input commitment *ic* is respected in C_n iff either:

- $ic \in C_n$, or
- $ic \notin C_n$, but there is a plausible argument for why a current commitment $c \neq ic$ should replace *ic*,[2] or
- $ic \notin C_n$, and there is a plausible argument for why *ic* does not belong to the subject matter (i.e., for why it is not relevant whether or not the target system can account for it).[3]

---

[1] I.e., it is possible to trade off independent credibility in order to increase account, as long as it can be plausibly argued that this independent credibility is outweighed. But it is not allowed to trade off *respect* for independent credibility in order to increase account.

[2] *c* can be a result of adjusting *ic* if $ic = \neg c$, but this is not necessary: It can also be that *ic* and *c* are consistent, but *ic* was rejected and *c* adopted instead because the candidate system S_n can account for *c* but is only consistent with *ic*.

[3] By allowing that commitments can be respected and be excluded from the subject matter at the same time, the respecting condition fits with Carnap's (later, pragmatic) understanding of the criterion of similarity for explications, which requires that the explicatum can be used instead of the explicandum in all relevant contexts (Brun 2020, 933)—i.e., we would explain here that this is *not* a relevant context. This, however, means that what relevant contexts are is not necessarily fixed from the beginning.

Whether or not the independent credibility of an input commitment is respected depends partly on how plausible the reasons are that can be given for its adjustment against the whole position (i.e., how well the system does justice to theoretical virtues, and how well commitments and system agree overall). This means that whether or not the independent credibility of a commitment *ic* is respected by, e.g., a specific commitment $c \neq ic$ (or the current set of commitments C_n as a whole), might change during the progress of the RE process and always has to be assessed anew.

Maybe it seems too strict to give lexical priority to the criterion of *Respecting Input Commitments* when adjusting commitments—maybe sometimes it will be necessary to explore various routes of adjustments before being able to vindicate the adjustment of a specific input commitment. But it is important to note that *commitment* refers to a specific epistemic state, i.e., not simply the content of a sentence, but being committed to what this sentence expresses. That an input commitment must not be adjusted until there is a plausible argument for this adjustment does not preclude the option of *tentatively* exploring what the consequences of adjusting this commitment would be for the position, and whether or not, looking back, we can defend adjusting the input commitment from the position that we ultimately reached. But until we can provide such an argument, the position will be "in the air", since we only tentatively try out what would happen if we were to adjust the input commitment, but without being able to defend said adjustment.

Doing Justice to Theoretical Virtues The target system should have theoretical virtues which can be measured at least on ordinal scales. The specific virtues and how they are measured and weighed is described in Sect. 5.5.

Having spelled out the criteria, let us now concretize the two steps of the process of adjustments:

Step A_{n+1}: Adjusting the System Adjust (a part of) the current system, S_n, with respect to (a subset of) the current commitments C_n. For this step, at least the following considerations are relevant:

(i) Assess and rank candidate systems with respect to how well they can account for current commitments C_n.

(ii) Assess and rank candidate systems with respect to their theoretical virtues.

(iii) Assess and rank candidate systems with respect to how well they can account for current commitments *and* do justice to theoretical virtues—ideally, a complete ordering will result; if not, describe the partial orderings that can be made and the trade-offs involved.

(iv) Based on the results of (i)–(iii), adopt a system S_{n+1} in order to continue the process, and provide reasons for why this candidate was chosen: If a candidate is pareto optimal with respect to account and the theoretical virtues, it has to be chosen. If no such candidate is available and trade-offs have to be made, they have to be defensible with respect to (a) their effects on the position as a whole, and (b) the pragmatic-epistemic objective.

Step B_{n+1}: Adjusting Commitments Adjust (a subset of) the current commitments C_n with respect to (a part of) the current system S_{n+1} by using the following strategy:

(i) For each commitment that is not fully accounted for, is there a way to adjust it in order to increase account that fulfills the respecting-condition? If yes, adjust, otherwise keep the commitment.[4]
(ii) Check whether all previous adjustments of input commitments still meet the respecting-condition. If not: Replace it by a commitment that does respect the independent credibility of the input commitment (this can also be the original input commitment).
(iii) Systematically explore: Are there further relevant commitments that, e.g., might conflict with the current system? If yes, add them to C_{n+1} (but do not yet adjust them[5]).
(iv) As a result of (i)–(iii), adopt a new set of current commitments C_{n+1} to continue the process, i.e., select a set of commitments that maximizes agreement with the current system S_{n+1} while respecting the independent credibility of input commitments.

As noted above, I decided not to explicitly include the assessment of support from background theories at each step. It can of course serve as tie-breaker in cases of trade-offs, and arguments referring to the background might often play a role when deciding between different possible adjustments. However, when the process of adjusting commitments and systems alternately comes to an end point—that is, when neither of the two steps leads to any further improvement of the position—we need to assess the resulting position with respect to all of the RE criteria, which includes the degree of support from background theories. As explained in Chap. 3, we then have to ask the following questions and assess to what degree the criteria are met:

- Are the resulting commitments and the system in agreement?
- Can the position be supported by background theories?
- Does the system do justice to theoretical virtues?
- When comparing input commitments and resulting commitments, is it plausible that we did not abandon the subject?
- Do the resulting commitments have independent credibility?
- Is the resulting position at least as plausible as relevant alternatives?

Having thus specified the method for its application, let us now turn to the description of the starting position, i.e., the input that we will work with.

---

[4] Commitments that were adopted at some point as, e.g., inferences from a system at an earlier state in the process and have no independent credibility: they can simply be adjusted, eliminated or replaced without further arguments. Commitments that got adopted as the result of adjusting an input commitment: When adjusting them, it still has to be defensible that the independent credibility of the original input commitment is respected.

[5] Since these emerging commitments did not play a role when selecting the current system, they should not at this step be adjusted with respect to it.

5.3 Initial Input Commitments

In line with what was said in Chap. 3, I made a selection of initial input commitments that I deem representative, or at least representative enough to start the process of adjustments with them. The initial commitments are the subset of the input commitments that enters the RE process as explicit input in the first step. The input commitments constrain the subject matter since they are what the target system has to respect.[6] Consequently, it makes sense that they do not only consist of case-specific, particular judgments that we are committed to. On the contrary, as the debate about precautionary principles and precaution (see Chap. 4) shows, we often seem to be quite confident about general statements like "uncertainty should not be a reason for inaction in the face of severe harm", or "the environment should be protected from serious irreversible harm, even if it is not certain that this harm would occur"—but will be less confident when it comes to deciding what the actual consequences are for individual decisions: if an action against a specific uncertain harm is very costly, which we know for sure, should we still take it? If a genetically modified crop could be used to avert an impeding hunger catastrophe, but there is a chance that its use will also have irreversible negative effects, e.g., on biodiversity, should we avoid using it?

As there are lots of different cases for which precaution is relevant, I decided to use as an example the case of the climate engineering strategy of solar radiation management (SRM) through stratospheric aerosol injections (SAI), so as to gain some focus for my selection of commitments. To make the selected commitments comprehensible, we need some background on this climate engineering strategy, which I describe in the following, before listing some examples of the selected commitments. Thus, what follows now is technically part of the background, which I only address in the next Sect. 5.4, but we need this information now in order to be able to correctly interpret some of the commitments.

5.3.1 An Illustrative Example: Precautionary Principles and Solar Radiation Management

Climate change is often cited as one of the paradigm case where a precautionary principle should apply (see, e.g., Gardiner 2006). Especially alarming is the possibility of so-called "runaway climate change", or "climate emergencies" (Blackstock et al. 2009). This refers to the possibility of passing certain thresholds that might accelerate climate change dramatically. Because of the possibilities for climate

[6] But they do not exhaust the subject matter, one reason being that a selection has to be made in order to keep the process manageable. Another reason is that for this case study, I only take commitments as input that can be expressed in the form of sentences—thereby excluding, e.g., commitments that are expressed in graphics, in actions, etc.

emergencies, even radical mitigation and adaptation measures might not be enough to avert catastrophic climate change impacts.

As a reaction, a range of technological approaches to alleviate the causes and/or effects of climate change have been suggested under the label of "geoengineering" or "climate engineering". Typically, a distinction between so-called "carbon dioxide removal (CDR)" and "solar radiation management (SRM)" strategies is made. CDR strategies aim at removing greenhouse gases from the atmosphere, e.g., via technological means or also via more "traditional" means like reforestation. SRM refers to technologies and measures with the goal to reduce global warming by enhancing the reflectivity of the earth. Examples range from painting roofs white to putting reflective aerosols in the stratosphere or even space mirrors. Since the proposed measures differ widely with respect to, e.g., their scale, costliness, speed of bringing about sizeable effects, associated uncertainties and possible side-effects, relating a PP to climate engineering in general is difficult (Elliott 2010).

Thus, I will specifically focus on large-scale SRM measures, i.e., stratospheric aerosol injections (SAI). While it is expected that this kind of SRM could cool the earth rapidly and cancel increases in global average temperature caused by high concentrations of greenhouse gases (GHGs), it would not compensate for other impacts of high levels of atmospheric GHG concentrations, e.g. ocean acidification. Also, impacts at a regional level and on other climate parameters are uncertain, e.g., how it will affect (regional) precipitation, and atmospheric and oceanic circulation. And the potential for so-called "unknown unknowns", i.e., completely unanticipated outcomes is high. Moreover, the research needed to potentially reduce uncertainties is itself beset with uncertainties and introduces new risks of its own (this description of SRM-SAI is based on Blackstock et al. 2009). Additionally, the so-called "termination problem" means that, as fast as SAI is expected to cancel out the warming from increased GHG concentrations, temperature would increase again equally quickly if we stopped engineering the climate abruptly. Especially if SAI would have been implemented without additional strict mitigation and adaptation measures, this increase could be devastatingly steep.

SRM-SAI and Precaution Expecting guidance from a precautionary principle with respect to the question of whether or not the solar radiation management (SRM) technology of stratospheric aerosol injections (SAI) should be researched and eventually deployed seems reasonable. Uncertainties are huge: We can identify outcomes that are seen as possible, but no or no reliable probability information is available, while it is plausible that there are even more possible outcomes that we haven't even been able to identify yet; and impacts (both of climate change without SAI, and of SAI itself) could be catastrophic from a human perspective.

Yet we can find arguments that invoke precaution on both sides of the debate: On the one hand, the "Lesser-evil argument" states that, because at a future point in time, SAI could be the lesser evil as compared to climate change catastrophes, we should research it now so that it is available then. It has been argued that research into SAI is in itself valuable, because it gives us an additional option, independently of whether or not we will make use of it (Reynolds and Fleurke 2013, 103). On the other hand, the "It-might-get-worse argument" states that SAI could, in the worst case, even worsen climate change catastrophes, and should, as a precaution, never be deployed and consequently shouldn't be researched either (for a reconstruction of the debate, see Betz and Cacean 2012). Additionally, there are concerns that SAI research could create some sort of "lock in" effect, leading via a slippery slope to the deployment of SAI even if no real catastrophe is impending, or that it could sideline the discussion and development of alternative approaches to deal with the threat of dangerous climate change (e.g., Fragnière and Gardiner 2016).

In part, these contradictory invocations of precaution rest on different empirical assumptions, e.g., about the possible side-effects or the psychological implications of SAI-SRM research. In part, they rest on different value bases, i.e., on a disagreement about *what* exactly the harms are that we should take precautions against, and what exactly makes them harmful. But to a large part, they also rest on a disagreement about what precaution means in this context, and on a lack of agreement on which precautionary principle to adopt (Elliott 2010).

For the case study, this poses the challenges of (i) how to assess the empirical background information and the scientific knowledge about effects and side-effects of SAI, (ii) how to evaluate different possible options and outcomes, and (iii) what the relevant factors are that a PP should take into account, and how it can guide us with respect to the results from (i) and (ii). Since I am interested in formulating an action-guiding PP that is part of a position in reflective equilibrium, my main goal is to address (iii). This presupposes answers to (i) and (ii), which would require doing a lot of further work before starting to tackle the problem of justifying a PP. I will therefore work as far as possible with plausible stipulations and assumptions with respect to (i) and (ii). The goal is to formulate a PP that can be coherently applied in a consistent framing of a decision-problem, and not to identify what the correct framing of a problem such as whether or not to research SAI is. Thus, throughout the case study, I will mostly work with what I call "toy examples", simplified case-descriptions of specific decision-problems.

5.3.2 Examples of Selected Commitments

For the purpose of this case study, I am only working with my own commitments (or rather with the commitments of a hypothetical person that is rather similar to me), even though that does not exclude adopting commitments based on arguments of others, e.g., convincing arguments from the literature. Based on, e.g., arguments that can be made in their favor independently of the current position, intuitions that are in line with a commitment, or also knowledge about how widely shared a commitment is, rough weights of *low–medium–high* are assigned to the commitments. When thinking about possible adjustments, these weights do not replace the need to consider the reasons for and against a commitment in detail. They only serve as a rough indication of the independent credibility of a commitment. As explained in Sect. 5.2, this ranking is only ordinal, i.e., it does expresses neither that two commitments with a high weight necessarily have the exact same degree of independent credibility, nor that there is a specific interval between the three weights.

I selected my initial commitments from the three groups of (1) general commitments about precaution and precautionary principles, (2) commitments to judgments in simplified "toy" examples, (3) commitments concerning an actual and complex problem, namely, research and development of solar radiation management (SRM) (although I will also use some toy examples for SRM in order to hopefully single out important aspects). As it is not possible to consider each and every one of my commitments concerning precaution and precautionary decision-making explicitly, I aimed for a representative selection of initial commitments. Still, throughout the process of adjustments, it will be important to search for further emerging input commitments which might be relevant.

A full list of the selected initial input commitments can be found in the Appendix A at the end of the book. In the following, I name some examples for each category, together with the rough weights of high, medium, and low assigned to them.

General Commitments About Precaution and Precautionary Principles Some of the general commitments are statements about, e.g., general features of situations that warrant precautions. Others put more direct demands, or constraints, on the target system, as they express commitments concerning what the target system should achieve, or what it could look like. The latter function as a sort of "working hypotheses", but are also commitments to features or functions/roles I expect the target system to have or fulfill. They can be adjusted or rejected just like every other commitment, e.g., by giving an argument for why this specific demand is not reasonable or cannot be consistently implemented—or maybe by showing that a system that does not meet these expectations does a better job of fulfilling the pragmatic-epistemic objective. Here are some examples:

IC 1 Where there are threats of serious or irreversible damage, lack of full scientific certainty shall not be used as a reason for postponing cost-effective measures to prevent environmental degradation. (Principle 15 of the Rio Declaration) [low]

IC 2 When an activity raises threats of harm to human health or the environment, precautionary measures should be taken even if some cause-and-effect relationships are not fully established scientifically. (Wingspread Formulation of the Precautionary Principle) [low]

IC 3 *Pro tanto*, it is better to take precautionary measures now than to deal with serious harms to the environment or human health later on. [high]

IC 6 If we are not sure whether a substance or technology is safe, but have a viable alternative that can be shown to be safe (at least with higher certainty than the option in question), we should use the alternative, even if it might be more costly in economic terms. [high]

IC 8 The structure of a PP includes two "trigger conditions", threat and knowledge, and a precautionary response. [low]

A low weight was assigned to IC 1 and IC 2 for the following reasons: they should be included and are an important part of the subject matter—indeed, they are often cited as paradigm examples. However, they are also often cited by critics who attack PPs for being vacuous or paralyzing, and there are good reasons to think that the target PP should differ from those two paradigm examples. IC 8 is widely endorsed in the literature, but as it is primarily concerned with the structure of a PP, it should only serve as a working hypothesis that can easily be given up. Consequently, I assign a low weight to it. IC 3 and IC 6, however, express what I take to be important and substantial claims about a PP. I thus assign a high weight to them.

Commitments About Toy Examples Commitments about toy examples are typically my own, intuitive judgements, and the weight indicates how secure I feel in this judgement. Some of the commitments about toy examples directly include the description of the case in question, like IC 14:

IC 14 You find a firearm, and from examining it, you come to the conclusion that it is not loaded. But you are aware that you don't know much about weapons—this is, in fact, the first firearm you have ever held in your hands. You must not point it at someone else and pull the trigger. Neither should you do the same with yourself. [high]

In other cases, the toy example is a bit more complex and the case description is separately listed as a part of the background. For example, take case 5, Asbestos 1:

Case 5: Asbestos 1 Large-scale mining and manufacturing of asbestos has started about 15 years ago. Asbestos is seen as a desirable material because of its properties like sound absorption, tensile strength, and its resistance to fire and heat. Production costs are low, so it is also affordable. However, there are observations and reports that associate lung diseases with inhaling asbestos, although no systematic scientific research has been done on it so far; thus, a clear connection cannot be proved, and the diseases might have other causes.

We have to choose between the following four options:

(i) BAU: Continuing business-as-usual,

(ii) Research: Starting systematic scientific research on the harmfulness of asbestos dust, including long-term studies and mortality statistics of asbestos workers,

(iii) Research&Regulation: Starting systematic scientific research while already strictly regulating asbestos production, including, e.g., limiting exposure of workers to asbestos dust, and making compensation arrangements, based on agreed liabilities, or

(iv) Ban: Banning asbestos.

Concerning this toy example, which is a simplified case based on real past events (cf. Harremoës et al. 2001), I have the following input commitment:

IC 15 In case 5, *Asbestos 1*, we should choose option (iii), Research&Regulation. [medium]

Some of the toy examples are also examples that I took from the literature, like case 12, Chemical Waste:

Case 12: Chemical Waste "A company applies for an emission permit to discharge its chemical waste into an adjacent, previously unpolluted lake. The waste in question has no known ecotoxic effects. A local environmental group opposes the application, claiming that the substance may have unknown deleterious effects on organisms in the lake.

[. . .] We know from experience that chemicals can harm life in a lake, but we have no correspondingly credible reasons to believe that a chemical can improve the ecological situation in a lake. (To the extent that this "can" happen, it does so in a much weaker sense of "can" than that of the original argument [. . .])." (Hansson 2016, 96)

Concerning this case, I have the following input commitment:

IC 22 In case 12, Chemical Waste, the company should not be allowed to discharge the chemical waste into the lake (example from Hansson 2016, 96). [high]

Commitments About R&D of Solar Radiation Management (SRM) These commitments refer to the case of solar radiation management, which was chosen as an illustrative example for the case study. The weights of these commitments mainly express how secure I feel in my judgement.

IC 23 Independently of whether or not SRM should be considered as part of precautionary measures in case the globally implemented mitigation and adaptation strategies turn out to be insufficient to prevent dangerous climate change, it should not be used as the only precautionary measure. ("SRM" here is short for "research and development on solar radiation management in order to have it ready to use should dangerous climate change be imminent".) [high]

IC 25 Non-invasive research into SRM should be done, as long as this does not negatively interfere with the search for and discussion of other approaches. [medium]

IC 26 A necessary condition for any application of SRM against harmful impacts of climate change is that it has to be accompanied by a strict mitigation and adaptation program that would allow us to stop doing SRM again as soon as possible. [medium]

IC 29 In case 3, *R&D into SRM, two kinds of research*, we should choose option (ii), doing non-invasive research into SRM, especially the aspects that contribute to our general understanding of climate science. [medium]

The last commitment in this list also refers to a toy example, which is described in case 2:

Case 2: R&D into SRM A strict mitigation and adaptation policy is implemented, but dangerous climate change is still possible because of feedback effects and tipping points. There are no signs that a catastrophe is imminent in the next 5 years. The basic mechanisms of solar radiation management are known, but there are still huge uncertainties, e.g. about its effects on a local level and possible (so far unforeseen, possibly catastrophic) side-effects. Should we do research and development on SRM with the goal of developing it ready to use?

We know that:[7] R&D has no, neither positive nor negative, influences on our mitigation and adaptation efforts, and that R&D itself does not pose any additional threats to the climate system.

We are given two choices: (i) implementing a research and development (R&D) program for SRM with the objective of developing SRM ready to use, or (ii) not implementing an R&D program for SRM.

5.4 The Background

In this section, I set out some parts of the background that likely will be relevant for the RE process, but this is not an exhaustive description. Also, I make some stipulations in the background in order to facilitate the case study.

Background Theories that might be relevant to argue for or against parts of the position are, e.g., rational choice theory, cost-benefit analysis, and maximizing expected utility theory.

Background Information is necessary in order to understand commitments, and to relate candidate systems to them. Concerning the case of solar radiation management, some background information is described above in Sect. 5.3.

For toy examples, I typically summarize the relevant background information in case descriptions like the one of case 5, Asbestos 1, which is also quoted above. Working with such toy examples allows me to have a simplified description of all the background information that is potentially relevant when it comes to assessing whether or not a given system is in agreement with a commitment. The full list of these case descriptions can be found in Appendix A at the end of the book.

Additionally, relevant background information includes knowledge about historical cases that are relevant for precautions, like the ones described in the case studies of "Late Lessons from Early Warnings" (Harremoës et al. 2001). For background information on current risk regulation practices, I use the description and discussion

[7] Obviously, these are simplifications which are not realistic.

in Randall (2011). As background information for solar radiation management, I use a narrow selection of relevant papers, especially Blackstock et al. (2009), Irvine et al. (2016), and Lenferna et al. (2017).

In order not to lose sight of the main line of the case study, I also work with a number of **background assumptions** and **stipulations**. For example, I assume that decisions concerning climate change policies take place under conditions of uncertainty, i.e., that no reliable probabilities about possible outcomes are available (Aldred 2013, 133). Additionally, I sometimes work with assumptions like stipulating numerical utilities, or stipulating that outcomes are in the relevant sense "reasonable" or "realistic"—i.e., I stipulate background information that is necessary to relate a candidate system to commitments. The results of the RE process are then of course contingent on whether or not the kind of information that I stipulate is actually obtainable in the real world.[8]

5.5 Theoretical Virtues

I am looking for a moral system—a moral precautionary *principle*, to be more precise—and consequently, the theoretical virtues that are relevant in this RE process should be relevant for moral theories. In the literature, we can find the following examples of virtues, or desiderata, for moral theories:

Determinacy: "A moral theory should feature principles which, together with relevant factual information, yield determinate moral verdicts about the morality of actions, persons, and other objects of evaluation in a wide range of cases" (Timmons 2012, 13).

Applicability: "The principles of a moral theory should be applicable in the sense that they specify relevant information about actions and other items of evaluation that human beings can typically obtain and use to arrive at moral verdicts on the basis of those principles" (Timmons 2012, 13).

Explanatory Power: "A moral theory should feature principles that explain our more specific considered moral beliefs, thus helping us understand why actions, persons, and other objects or moral evaluation are right or wrong, good or bad, have or lack moral worth" (Timmons 2012, 15); "A theory has explanatory power when it provides enough insight to help us understand the moral life: its purpose, its objective or subjective status, how rights are related to obligations, and the like" (Beauchamp and Childress 2013, 340); "Moral theories should identify a fundamental principle that both (a) explains why our more specific considered

[8] For example, if a position is brought into reflective equilibrium contingent on the assumption that we can assign numerical utilities to outcomes, then it is not necessary that the same utilities can be assigned, just that it is possible in general. If it turns out that this stipulation does not correspond to reality, then the justification collapses.

moral convictions are correct and (b) justifies them from an impartial point of view" (Hooker 2000, 4).

Clarity: "A theory should be as clear as possible, as a whole and in its parts" (Beauchamp and Childress 2013, 339).

Completeness and Comprehensiveness: "A theory should be as complete and comprehensive as possible. A theory would be fully comprehensive if it could account for all moral values and judgments. Any theory that includes fewer moral values will fall somewhere on a continuum, from partially complete to empty of important values" (Beauchamp and Childress 2013, 339).

Simplicity: "A theory should have no more [basic] norms than are necessary, and no more than people can use without confusion" (Beauchamp and Childress 2013, 339).

Output Power: "A theory has output power when it produces judgments that were not in the original data base of particular and general considered judgments on which the theory was constructed. If a normative theory did no more than repeat the list of judgments thought to be sound prior to the construction of the theory, it would have accomplished nothing. For example, if the parts of a theory pertaining to obligations of beneficence do not yield new judgments about role obligations of care in medicine beyond those assumed in constructing the theory, the theory will amount to no more than a classificatory scheme. A theory, then, must generate more than a list of the axioms already present in pretheoretic belief" (Beauchamp and Childress 2013, 340); "Moral theories should help us deal with moral questions about which we are not confident, or do not agree" (Hooker 2000, 4).

Practicability: "A proposed moral theory is unacceptable if its requirements are so demanding that they probably cannot be satisfied or could be satisfied by only a few extraordinary persons or communities. A moral theory that presents utopian ideals or unfeasible recommendations fails the criterion of practicability" (Beauchamp and Childress 2013, 340).

Based on this list, I selected Practicability, Determinacy, Broad Scope, and Simplicity as theoretical virtues for my RE project; although my understanding of them often differs from the one mentioned above. (And Scope is missing from the above list altogether.) In the following, I detail my understanding of these virtues, and give some reasons for why I selected them.

I decided to leave out "explanatory power" since it is notoriously unclear what this exactly means—and many attempts to explicate it ultimately refer to other theoretical virtues (cf. Keas 2017; Ylikoski and Kuorikoski 2010).

"Output Power", the demand that the system should not merely "repeat the list of judgments thought to be sound prior to the construction of the theory", and that it should "help us deal with moral questions about which we are not confident" is certainly desirable, but also seems rather have to do with the relation of the system to commitments, and its scope. Thus, I take it that this desideratum is already covered by other aspects of RE.

I subsume "Applicability" under **Practicability**, which I understand as having to do with whether or not we can apply the system and follow its instructions—e.g., how accessible the necessary information is typically for us. As Roser (2017, 1397) argues: in order to be action-guiding, a principle "must process inputs that are available to us". Consequently, it should also produce outputs that are implementable/realizable for us, respectively identifiable by us. E.g., a target system that tells us to select the course of action that will bring about the least actual harm is not very applicable if we want to know what to do in a situation that is exactly characterized by the fact that we do not know which course of action will effectively lead to how much harm.

On the other hand, such a system still can have **Determinacy**: "The course of action that will bring about the least actual harm" is not a vague expression, and if the necessary information were accessible to us, we would have no problem in applying it and identifying the correct course of action. I understand "Determinacy" as the opposite of vagueness and imprecision, as requiring that the conditions of application of a system are precise and clear enough to yield, together with relevant background information, definite verdicts. E.g., it should—given relevant background information—allow us to unequivocally identify whether a specific course of action is permissible, required, or prohibited. For example, a system that tells us to choose the most sustainable course of action is not very determinate if it does not also specify what makes a course of action "the most sustainable one".

The virtue of determinacy is, to some extent, dependent on background information: if we have, e.g., a suitable explication of "sustainability" already in our background, then a system like my last example will be determinate without having to specify sustainability. Similarly, it could be the case that we only have a very vague or imprecise concept of "harm", making the first example less determinate.

One can ask whether it also could be that the system is not determinate because its conditions of application and/or its verdicts are too general. But generality is not a problem in itself, as long as every case that falls under the general conditions is truly a case where the principle should apply. When the generality of the antecedent is a problem, this is rather because it is not clear whether a given case falls under it (but this is a problem of vagueness or ambiguity), or because there is a case that falls under it but we feel that it should not (and this is a problem with account, but not determinacy). If these problems do not occur, then generality is even desirable on grounds of the theoretical virtue of *broad scope*: a more general antecedent will typically be applicable to a broader range of cases than one that is more specific.

"Comprehensiveness and Completeness" is already partly covered by the RE criterion of *Account*. But there is something virtuous about the range of applicability of a system that is not completely covered by its ability to account for commitments: we want a system that is applicable to a broad range of cases, i.e., we want a system to have a **Broad Scope**. While scope is connected to account, it is not the same: *Account* is about the relation between commitments and the target system, i.e., demanding that the target system can account for the commitments. This means that account can also be increased by rejecting commitments the target system cannot account for, or by excluding them from the subject matter, e.g., by continuing to be committed to something but realizing that it is not relevant for *precaution*. While these strategies would increase account, they would at the same time reduce the scope of the target system.

I understand the virtue of scope in terms of the *range of applicability* of a system. My use of "range of applicability" is based on Scharp (2013, 40), although it has to be adapted from the applicability of concepts (in Scharp's case) to the applicability of systems (in the RE context). I take it that the range of applicability of a system consists of those classes of facts in which the system yields a verdict, which can mean that it prescribes or prohibits an action, but also includes those cases in which it tells us that a specific action is permissible, or not required. I.e., for a precautionary principle, its range of applicability consists of all those situations in which it tells us whether or not a (specific) precautionary measure should be taken. Continuing the adaptation of Scharp's terminology, the *application set* of a PP would then consist of all those cases in which it prescribes precaution, and its *disapplication set* all those cases in which it tells us that no precaution is required.[9] But this latter distinction is less relevant, at least for the virtue of scope: while we want a PP with a broad range of applicability, i.e., a PP that tells us in as many cases as possible whether or not we should take precautions, this does not mean that we are looking for one that prescribes precautionary measures in as many cases as possible.[10]

This means, e.g., that a target system that includes necessary and sufficient conditions for precaution has a broader scope than if the same conditions were only sufficient. Take two principles, P_1 and P_2: P_1 has the form "If conditions *a* and *b*, then take measure *c*". This means that it is applicable to all cases that have the properties *a* and *b*. If one of the two is missing, P_1 is not applicable. Since *a* and *b* are sufficient but not necessary conditions to take measure *c*, we do not know what to do if one of them is missing: both that *c* is permissible or that it is prohibited would be consistent with P_1 in such a situation. For P_1, its range of applicability coincides with its application set.

On the other hand, P_2 has the form "If and only if conditions *a* and *b*, take measure *c*". Since here, *a* and *b* are not only sufficient but also necessary conditions for *c*, P_2 is thus applicable to the classes of cases and situations that have the properties (*a*, *b*), (*a*, *non-b*), (*non-a*, *b*), and (*non-a*, *non-b*): in all these combinations, it tells us whether or not we should take measure *c*. Its range of applicability is thus broader than its application set, i.e., the situations in which its conditions are fulfilled and it prescribes *c*—and also broader than the one of P_1.

Notably, the virtue of *Broad Scope* has to do with the range of classes of situations to which a system is applicable. This does not say anything about, e.g., how prevalent these classes are, or whether or not the most relevant commitments

---

[9] Thus, a minimal requirement for a system, which has more to do with account than with scope, is that in order to avoid inconsistencies, (i) the range of applicability and the range of disapplicability of a system have to be disjoint, and (ii) the application set and the disapplication set of a system have to be disjoint (Scharp 2013, 40–41).

[10] This conception of "scope" is clearly oriented towards action-guiding (moral) principles, which is my focus here. But it should be possible to adapt it also for other contexts.

belong to them. Whether the range of applicability of a system actually covers the *relevant* cases is, on my understanding, not a question of broad scope, but of assessing account for (weighted) commitments.

Simplicity is one of the "classic" theoretical virtues, but this does not mean that there is one straightforward interpretation of it.[11] In the context of this project, I settled for an interpretation of the theoretical virtue of simplicity as demanding that the conceptual apparatus of the target system should be economical in the sense that the concepts it includes that cannot be reduced to each other are kept to a minimum. There might not be a direct argument for why simplicity, understood in this sense, has epistemic virtue. But striving for simplicity will contribute to systematicity since it forces us to, e.g., identify relevant features that commitments *share* in order to reduce the number of concepts needed to account for them.[12]

Operationalization and Weighing of Virtues Here is a list summarizing the theoretical virtues of systems that I selected for the case study:

Practicability The target system should be applicable in the sense that it specifies relevant information about actions and other items of evaluation that human beings can typically obtain and use to arrive at moral verdicts. E.g., it should process inputs that are typically available to us, and yield verdicts that are realizable by us.

Determinacy The target system should, together with relevant factual information, yield determinate verdicts, i.e., both its conditions of application and its verdicts should be precise and clear enough.

Broad Scope The target system should have a broad range of applicability, i.e., it should tell us in as many cases as possible whether or not (and specifically which) precautionary measures are required.

---

[11] According to Kuhn (1977, 322), simplicity means that a theory should bring "order to phenomena that in its absence would be individually isolated and, as a set, confused". As some further interpretations of simplicity as a theoretical virtue, Lacey (2005, 60) names *parsimony*; *economy* (of formulation, of technical devices); efficiency in use for explanatory, predictive and other "scientific" purposes; deployment of the "simplest" available mathematical equations; conceptual clarity; *idealization* which provides a benchmark, departures from which can be conveniently explained; and *formalizability*. Simplicity is often seen as being closely connected to unification, e.g., that it is desirable for a theory to be simple because systematizing sets of data/commitments with a smaller theoretical apparatus will serve to make patterns explicit and thereby increase our understanding of the domain in question (e.g., Sachs 2017, 28). Another related understanding is that of a simple theory as one that is in some sense "easy to grasp" (Beauchamp and Childress 2013, 339), which includes that a theory should have no more norms "than people can use without confusion".

[12] This account of simplicity is loosely based on the one of Goodman (e.g., Goodman 1943, 1955), although simplified (no pun intended). For the purpose of the case study, using an axiomatic like the one proposed by Goodman seems too demanding, and I expect that this conception of simplicity will fulfill the function of allowing me to assess and compare competing candidate systems.

Simplicity The conceptual apparatus of the target system should be economical in the sense that the concepts it includes that cannot be reduced to each other are kept to a minimum.

I selected this specific list of virtues of the system because I take them to be relevant to reach my pragmatic-epistemic objective of formulating a defensible, action-guiding moral principle that is applicable to the subject matter of precaution and precautionary decision-making. As I argued in Chap. 3, there is no unequivocal ranking of virtues in case of trade-offs: rather, such trade-offs have to be decided on a case-by-case basis, and need to be defensible with respect to (a) the pragmatic-epistemic objective, and (b) their effects on the position as a whole.

Thus, while I expect that practicability and determinacy will typically be more important than scope in trade-offs, whereas simplicity might be not much more than a "tie-breaker" when candidate systems are otherwise equally virtuous, this will always have to be defended on a case-by-case basis.

5.6 Candidates for the System

As a result of the literature survey on precautionary principles in Chap. 4, the Rawlsian Core Precautionary Principle (RCPP) of Gardiner (2006), the integrated risk regulation framework of Randall (2011), the tripartite proposal of Steel (2015), and the Catastrophic Harm Precautionary Principle of Hartzell-Nichols (2017) were identified as promising candidates for a PP. I do not explicitly consider the proposals of Randall and Hartzell-Nichols, for the following reasons: the Catastrophic Harm PP of Hartzell-Nichols comes with a framework that includes substantial procedural aspects which are difficult to simulate in an RE process that is conducted by a single epistemic agent like myself. The proposal of Randall has a strong focus on risk assessment and risk management, which provides relevant background information, but does not seem very promising as a candidate for a *moral* precautionary principle. This leaves us with the RCPP of Gardiner (2006) and the tripartite approach of Steel (2015). I will also consider Bognar's counterproposal against the RCPP. Bognar (2011) argues that a "utilitarian principle", which consists of a combination of the principles of indifference and of maximizing expected utility, can equally well account for the cases where the RCPP applies, while having a broader scope.

In addition to critically assessing and comparing these candidates in the process of adjustments, I will also explore how one can develop new candidates for the system as part of the process of adjustments.

5.7 Recapitulation: Design of the Case Study

This chapter has first specified the method of RE for its application to the project of justifying a moral precautionary principle. It then identified the various elements that enter the process of adjustments—initial commitments, elements of the background, theoretical virtues, and candidates for the system. I named some examples of the selected initial commitments, and the full list can be found in the appendix at the end of the book. The appendix plays an important role for the case study, as when applying RE throughout Chaps. 6–8, I will continue to only discuss representative or relevant aspects.

RE will be applied with the pragmatic-epistemic objective of justifying an action-guiding moral principle that is applicable to the subject matter of precaution and precautionary decision-making. At the same time, I have the goal of putting the method itself to a test: the case study will demonstrate how the method of RE can be applied, and test the way the method was conceptualized in Chaps. 2 and 3. To be able to focus in detail on specific aspects of applying RE, the case study is divided into three phases: Phase 1 (Chap. 6) explores how theory construction, i.e., the development of a candidate system, works in RE. Phase 2 focuses in detail on the two steps of the process of adjustments (Chap. 7). And the third phase works towards a preliminary end point of the RE process, and evaluates this resulting position (Chap. 8).

References

Aldred J (2013) Justifying precautionary policies: incommensurability and uncertainty. Ecol. Econ. 96:132–140. https://doi.org/10.1016/j.ecolecon.2013.10.006

Beauchamp TL, Childress JF (2013) Principles of biomedical ethics, 7th edn. Oxford University Press, Oxford

Betz G, Cacean S (2012) Ethical aspects of climate engineering. KIT Scientific Publishing, Hoboken

Blackstock JJ, Battisti D, Caldeira K, Eardley DM, Katz J, Keith WD, Patrinos AAN, Schrag DP, Socolow RH, Koonin SE (2009) Climate Engineering Responses to Climate Emergencies. Technical report, Novim

Bognar G (2011) Can the maximin principle serve as a basis for climate change policy? Monist 94(3):329–348. https://doi.org/10.5840/monist201194317

Brun G (2020) Conceptual re-engineering: from explication to reflective equilibrium. Synthese 197(3):925–954. https://doi.org/10.1007/s11229-017-1596-4

Elliott K (2010) Geoengineering and the precautionary principle. Int J Appl Philos 24(2):237–253

Fragnière A, Gardiner SM (2016) Why Geoengineering Is not "Plan B". In: Climate justice and geoengineering: ethics and policy in the atmospheric anthropocene, p 15

Gardiner SM (2006) A core precautionary principle. J Polit Philos 14(1):33–60

Goodman N (1943) On the simplicity of ideas. J Symb Log 8(4):107–121

Goodman N (1955) Axiomatic measurement of simplicity. J Philos 52(24):709–722. https://doi.org/10.2307/2022696

Hansson SO (2016) Evaluating the uncertainties. In: Hansson SO, Hirsch Hadorn G (eds) The argumentative turn in policy analysis. Springer, Berlin, pp 79–104

Harremoës P, Gee D, MacGarvin M, Stirling A, Keys J, Wynne B, Vaz SG (eds) (2001) Late lessons from early warnings: the precautionary principle 1896–2000. Office for Official Publications of the European Communities, Luxembourg
Hartzell-Nichols L (2017) A climate of risk: precautionary principles, catastrophes, and climate change. Routledge, New York
Hooker B (2000) Ideal code, real world: a rule-consequentialist theory of morality. Oxford University Press, Oxford
Irvine PJ, Kravitz B, Lawrence MG, Muri H (2016) An overview of the Earth system science of solar geoengineering. Wiley Interdiscip Rev Clim Chang 7(6):815–833. https://doi.org/10.1002/wcc.423
Keas MN (2017) Systematizing the theoretical virtues. Synthese 195(6):2761–2793. https://doi.org/10.1007/s11229-017-1355-6
Kuhn TS (1977) The Essential Tension. Selected studies in scientific tradition and change. The University of Chicago Press, Chicago
Lacey H (2005) Is science value free?: values and scientific understanding. Psychology Press, Hove
Lenferna GA, Russotto RD, Tan A, Gardiner SM, Ackerman TP (2017) Relevant climate response tests for stratospheric aerosol injection: A combined ethical and scientific analysis. Earth's Future 5(6):577–591. https://doi.org/10.1002/2016EF000504
Randall A (2011) Risk and Precaution. Cambridge University Press, New York
Reynolds JL, Fleurke F (2013) Climate engineering research: a precautionary response to climate change? Carbon and Climate Law Review 7(2):101–107
Roser D (2017) The irrelevance of the risk-uncertainty distinction. Sci Eng Ethics 23(5):1387–1407. https://doi.org/10.1007/s11948-017-9919-x
Sachs B (2017) Explaining right and wrong: a new moral pluralism and its implications. Routledge, London
Scharp K (2013) Replacing truth. Oxford University Press, Oxford
Steel D (2015) Philosophy and the precautionary principle. Cambridge University Press, Cambridge
Timmons M (2012) Moral theory: an introduction. Rowman and Littlefield Publishers, Lanham
Ylikoski P, Kuorikoski J (2010) Dissecting explanatory power. Philos. Stud. 148(2):201–219. https://doi.org/10.1007/s11098-008-9324-z

Chapter 6
Case Study, Phase I: Developing a Candidate System

In the following three Chaps. 6–8, reflective equilibrium, as described in Chaps. 2 and 3, is applied in a case study with the input and design as described in Chaps. 4 and 5. I start with an introduction to the case study as a whole, before giving an overview of the current chapter.

6.1 Overview: Three Phases of the Case Study

For this application of reflective equilibrium (RE), **two objectives** have to be distinguished: *In* the application of RE, I pursue the pragmatic-epistemic objective of justifying a moral precautionary principle (PP). But *with* this application, I pursue the goal of testing how RE can be implemented and what we can learn for the method of RE by putting it into practice.

In the first place, this is a case study for RE, meaning that the second objective takes precedence. However, the first objective is still very important, not least because one aspect of evaluating the applicability of RE (goal 2) will depend on how well the application of RE contributes to the goal of justifying a moral PP (goal 1). This means that even though not every detail of it will be spelled out, and I will sometimes work with plausible assumptions and stipulations, the precaution-content must not be oversimplified in order not to run counter to the goal of testing RE with respect to an actual and complex problem. Still, the focus of the case study is on aspects of the application that are interesting from the perspective of RE, that is, on exploring different ways to use the method, on how specific problems for its application can be solved, etc. To achieve this, the RE process is roughly divided into three parts:

Phase 1 is the content of the present Chap. 6. In it, I explore how theory *construction* works in RE, i.e., how first candidate systems can be developed starting from the initial commitments. I develop a first preliminary candidate and give an

T. Rechnitzer, *Applying Reflective Equilibrium*, Logic, Argumentation & Reasoning 27, https://doi.org/10.1007/978-3-031-04333-8_6

outlook on how it might be improved. But in order to also test other aspects of an RE implementation, I then move on to phase 2 instead of fully developing this candidate.

In phase 2, Chap. 7, the focus is on the steps of alternating between adjusting commitments and adjusting the system. While these steps are applied in all three chapters, Chap. 7 describes them in most detail, and uses them to assess and compare a range of candidate systems. In particular, it shows how the RE criteria—as specified in Chap. 5—can be used to assess possible adjustments to the position, and how trade-offs can be resolved. To have real variety in the compared systems, I adopt candidates for PPs from the literature (see Chap. 4) in order to compare them with RE, with respect to my input commitments and my pragmatic-epistemic objective.

In phase 3, Chap. 8, I focus on a specific pathway of the RE process and develop a rights-based PP in answer to the pragmatic-epistemic objective of my RE process, i.e., to justify a *moral* precautionary principle. Compared with phase 2, I move away from assessing many different candidate systems, and focus on making one candidate as strong as possible. At the end of this final phase, I show how the RE criteria can be used to assess whether or not a justified position in reflective equilibrium was reached.

These phases are not an inherent feature of RE. It is by no means necessary that a process will proceed in these three phases, or that we will always find them. Nonetheless, each of them focuses on relevant aspects an RE process can, and typically will, have: (a) the construction of a system, often involving sub-processes, e.g., explications, (b) the comparison of, and choice between, different possible adjustments, given a broad range of candidate systems, and (c) spelling out, assessing, and defending a particular position in detail. These aspects can also appear together, or in a different order. Dividing the case study in three phases, each focusing on one of these aspects, allows me to go into more detail with respect to each of them.

Gray Boxes This level of detail is necessary to really test the RE method and not to gloss over important challenges, but it can be difficult to keep track of the big picture. To help readers follow the process, gray boxes summarize the main points of each step. Readers not interested in every detail should be able to get a general idea of the case study by reading the gray boxes and then the recapitulation and discussion of results at the end of each chapter.

Role of the Appendix Even though the description is often very detailed, it is impossible to describe everything. Thus, only relevant or exemplary aspects of what I did when applying RE are described—for example, the assessment of account for commitments is exemplary for a small selection of commitments, but results for the set of commitments as a whole are only summarized. To describe how account was assessed for each individual candidate system with respect to each individual commitment etc. would take too much space and become excessively repetitive. For reference, all of the commitments, candidates for (parts of) the system, and background information can be found in Appendix A at the end of the book.

6.2 Overview: Phase 1

In the first phase, the focus is on how we can develop candidate systems within the RE framework. If no candidate is already available, one natural starting point is to survey one's commitments for suitable candidates. Thus, I start in Sect. 6.3 with an A-step, i.e., with adjusting the system, in the very specific sense of *constructing* a first system. I adopt two general commitments—the Rio and the Wingspread PP—as candidate systems, and assess them comparatively with respect to commitments and theoretical virtues. On this basis, I identify guiding questions for further system development. These guiding questions are used in Sect. 6.4 in a B-step to systematically broaden the set of commitments, which leads to the formulation of working hypotheses. These working hypotheses are weak emerging commitments to certain specifics of the target system, e.g., functions it should fulfill or elements we expect it to have. Thus, they are at the same time tentative attempts towards a systematization of the subject matter.

One problem of both candidate principles is that it is unclear what counts as a "precautionary measure". The system is thus further developed through a "sub-RE-process", in this case, an explication: the goal is to explicate a part of the system (the concept of "precautionary measures"). Consequently only this part, and the relevant subset of the commitments, are adjusted with respect to each other in steps A_2 and B_2, Sects. 6.5 and 6.6. This demonstrates how explications can, as "sub-RE-processes", be part of system development in RE. The resulting explication and some of the working hypotheses are then put together in order to formulate a first candidate system in Sect. 6.7. This formulation is a part of the next A-step, A_3, which will be continued in the next Chap. 7, when this candidate is assessed in comparison with other candidates.

Section 6.8 recapitulates the main results from the first phase and discusses some intermediate results both with respect to RE as well as with respect to PPs. It also gives a schematic summary of the RE steps from phase 1 in Fig. 6.1.

The description of this first phase is structured along the lines of the two RE steps. However, they are not fully completed, since, e.g., in step A_1, none of the two candidates is selected. And in step B_1, commitments are not adjusted with respect to a system, but rather with respect to guiding questions that are a result of the partial implementation of step A_1. However, even if the steps are not completely implemented in every respect, I argue that what is done can reasonably be seen as partial instances of them.

6.3 Step A_1: Assessing Rio and Wingspread as First Candidate Systems

In this step, two input commitments, the Rio and the Wingspread formulation of a precautionary principle (PP) are assessed as candidate systems. They are rejected as inadequate candidates: They attain a very low account value (Sect. 6.3.1) and have a low theoretical virtuousness (Sect. 6.3.2). None of them can defensibly be adopted, so no system is chosen at the end of this step. However, their assessment allows for the formulation of guiding questions toward the construction of a new candidate system (Sect. 6.3.3).

While we already identified some possible candidates for the system in the design of the case study (Chap. 5), the goal of this first phase is to explore how RE can be used to develop new candidates. So let us for the moment assume that we do not already have plausible candidates for the system. How could we proceed? As explained, one natural starting point would be to survey one's commitments for suitable candidates. Looking at the set of input commitments, almost any of the general commitments about Precaution and Precautionary Principles (see A.1.1.1) could be tested as candidate systems.[1] I decided as an example to focus on principle 15 of the Rio declaration, and the Wingspread formulation of a precautionary principle, which are typical starting point for discussions about PPs.

IC 1 Where there are threats of serious or irreversible damage, lack of full scientific certainty shall not be used as a reason for postponing cost-effective measures to prevent environmental degradation. (Principle 15 of the Rio Declaration) [low]

IC 2 When an activity raises threats of harm to human health or the environment, precautionary measures should be taken even if some cause-and-effect relationships are not fully established scientifically. (Wingspread Formulation of the Precautionary Principle) [low]

I am committed to those because they are often cited as paradigm examples of precautionary principles (Ahteensuu 2008, 79), and I think that because of this, they determine important aspects of the subject matter. However, I only assign a low weight to them, since they also face a lot of criticism and are typically the start, not the endpoint of attempts to formulate and defend a PP (compare the survey on PPs, Chap. 4). But what exactly are the problems with Rio and Wingspread? Assessing them as candidate systems according to the RE criteria allows us to systematically identify their weaknesses, i.e., it helps us to work towards developing stronger candidates. Thus, I adopt them as the following two candidate principles:

[1] Aside from IC 8, which is a commitment about the structure of a PP and would at least have to be supplemented with further information/principles before it could be applied to the subject matter.

Principle 1 (P 1, The Rio PP) Where there are threats of serious or irreversible damage, lack of full scientific certainty [about those threats, T.R.] shall not be used as a reason for postponing cost-effective measures to prevent environmental degradation.

Principle 2 (P 2, The Wingspread PP) When an activity raises threats of harm to human health or the environment, precautionary measures should be taken even if some cause-and-effect relationships are not fully established scientifically.

Even when adopting them as candidate systems, they remain in my set of current commitments and need to be accounted for. Also note that this does not mean that I am committed to them *as candidate systems*.

After assessing how well P 1 and P 2 can account for commitments (Sect. 6.3.1) and their theoretical virtues (Sect. 6.3.2), I formulate guiding questions for the further development of a new candidate system (Sect. 6.3.3).

6.3.1 Rio and Wingspread: Account for Commitments

Principle 1 and Principle 2 are virtually never able to account for commitments. There are some borderline cases where one could argue that the commitment is accounted for *if* we were to presuppose additional information, but no clear-cut case of account aside from the fact that, trivially, they account for themselves (because Rio and Wingspread are themselves commitments, they can of course account for themselves).

Assessing Account: Some Examples In the following, I use two commitments to exemplify how account was assessed, and to demonstrate how P 1 and P 2 relate to commitments, but fail to account for them.

First, here is a commitment concerning precaution and the climate engineering technology of solar radiation management (SRM):

IC 23 Independently of whether or not SRM should be considered as part of precautionary measures in case the globally implemented mitigation and adaptation strategies turn out to be insufficient to prevent dangerous climate change, it should not be used as the only precautionary measure. ("SRM" here is short for "research and development on solar radiation management in order to have it ready to use should dangerous climate change be imminent".) [high]

Principle 1, the Rio PP: Its conditions for application are met, but it cannot account for the commitment. There are "threats of serious or irreversible damage": (a) threats from dangerous climate change, e.g., it could be that global average temperature rises rapidly because of positive feedback loops, as well as (b) threats from solar radiation management, e.g., disruptions of local precipitation patterns that could have further disruptive effects on the global climate. We don't have "full scientific certainty" about either of these threats. Thus, P 1 tells us that this uncertainty cannot be used as a reason against taking measures to prevent environmental degradation that could be caused by those threats. But even if we assume that there are no other reasons against taking such measures, P 1 cannot account for the commitment: It would follow that uncertainty must not be a reason for postponing measures to prevent (a) and (b), but it does not follow that (b), i.e., SRM, cannot be the only measure against (a), as long as we also take measures against the threats of (b). P 1 is, however, at least consistent with IC 23, since neither does it follow that SRM should be the only precautionary measure against the threat of failed mitigation and adaptation strategies.

Principle 2, the Wingspread PP, also applies in the sense that its conditions for application are met, but it can't account for the commitment either. The reasoning is similar: Applying P 2 tells us that we should take precautionary measures against (a) the threats from dangerous climate change as well as against (b) the threats from researching, developing, and deploying SRM. But it does not tell us what kind of precautionary measures we should take, and consequently cannot tell us whether or not the combination of "SRM + precautionary measures against the threats of SRM" is on its own an adequate precautionary measure against threat (a).

As the second example, here is a commitment to a decision in a toy example:

IC 22 In case 12, Chemical Waste, the company should not be allowed to discharge the chemical waste into the lake (example from Hansson 2016, 96). [high]

Principle 1, the Rio PP, is at least consistent with this commitment. There is a threat of serious damage (deleterious effects on organisms in the lake). Arguably, not allowing the discharge of the waste is a cost-effective measure to prevent environmental degradation (e.g., more cost-effective than cleaning up the waste again in case there is indication that it actually causes harm). Thus, the lack of full scientific certainty about the threat shall not be used as a reason not to forbid the discharge of the waste. However, this does not amount to a full account of the commitment that the waste *should not* be discharged. There is additional information which allows us to construct an argument which accounts for the commitment, and which *includes* P 1: according to the background information, there are no other (relevant, important) reasons against prohibiting discharging the chemical waste, and there is already a *demand* to prohibit it. But P 1 does not on its own demand that the waste should not be discharged. It can at best partly account for the commitment.

Principle 2, the Wingspread PP, cannot account for the commitment. We could try to argue that, similar to the reasoning when applying P 1, we can assume that not allowing the discharge of the chemical waste is an adequate precautionary measure and therefore demanded by P 2. But it seems to me that, just as well, "precautionary measures" could mean that we have to take measures to monitor the lake in order to react quickly when there are indications of harm, etc. Thus, P 2 is only consistent with the commitment.

6.3.2 *Rio and Wingspread: Theoretical Virtues*

Although Principle 1 and Principle 2 have a low account value, their conditions of application are often approximated or even met. The reasons for this slightly surprising result—i.e., that the conditions of application are often approximated, yet the candidates fail to account for commitments—also has to do with their theoretical virtues.

The examples of IC 23 and IC 22 are characteristic of the failure of P 1 (the Rio PP) and P 2 (The Wingspread PP) to account for commitments. If we roughly assess the theoretical virtues of these two candidates, we can see some of the reasons for this failure: most strikingly, both candidates do exhibit a very low degree of Determinacy. One reason for this is that a number of concepts that are used in them could be interpreted in different ways, e.g., "lack of full scientific certainty", "cost-effective", "serious or irreversible damage", "not fully established scientifically", or "precautionary measures". In fact, without clarifying these concepts, it is not even really possible to assess the applicability of the two principles: maybe we could argue that, in general, we have some implicit, pre-theoretic understanding of what, e.g., "serious" damage is, or what the relevant sense of "irreversible" damage is, or what does or does not count as a "precautionary measure", and that there are cases where the relevant information is accessible to and processable by us. But given only these pre-theoretic, imprecise concepts, there will be many boundary cases where we will just not be able to understand what the principle even requires of us.

In short, even though there is currently no other candidate available, it seems that P 1 and P 2 should be dismissed. However, by analyzing their shortcomings, we are now in a position to identify questions that need to be addressed in order to formulate a more promising candidate system.

6.3.3 *Formulating Guiding Questions*

Based on the identified shortcomings of the Rio and the Wingspread PP, guiding questions for the further development of the system are formulated.

(continued)

These express desiderata for the system, i.e., the target system should be able to provide answers to them.

While both candidates were often applicable in a way that did not contradict the commitments, the results were typically too uninformative to account for the commitments.

So one important objective for further candidate principles is that they should yield more informative verdicts, e.g., with respect to the **characterization of precautionary measures**: while the Wingspread PP (P 2) does not characterize the required "precautionary measures" any further, the Rio PP (P 1) at least asks for "cost-effective measures to prevent environmental degradation". But this is still lacking in clarity: does "prevent" mean that the measures have to guarantee (to some sufficient degree) that harm can be avoided? Or would it be enough if at least part of the possible harm were prevented? In any case, it seems to exclude measures such as doing further research to get a better understanding of the threat from counting as precautionary measures, and while one could argue that such an exclusion is indeed reasonable, it does not seem to fit with what I am committed to.

Then there is the question of what exactly "cost-effective" means: taking the cheapest measures available? The ones that promise the greatest net benefit? The ones that promise to achieve a given goal in the least costly way?

Additionally, the Rio PP does not directly demand of us that we take precautionary action, but states that "lack of full scientific certainty *shall not be used as a reason for postponing* [italics T.R.]" measures. This means that in order for there to be a demand for action, (a) there has to be some *other* demand or imperative for action, and (b) we always have to consider whether there are valid other reasons not to take the measures. Clause (b) in itself is certainly reasonable, but the Rio PP gives us no guidance in determining what these other reasons could be. Because of (a), the Rio PP itself is not directly demanding action, but is rather an argumentative principle that says something about what kinds of arguments are admissible (Sandin et al. 2002).

Also, neither of the candidate principles takes into account that actions that pose a threat often also bring chances of benefits, and consequently they do not say anything about trade-offs, and could not help when choosing between different proposed measures that also introduce their own threats.

A further problem is that the **scope of application** of both principles is somewhat **unclear**: the Rio PP refers to "serious or irreversible" threats, but then measures should only be taken to protect the environment—this leads to the somewhat puzzling consequence that there might be threats where the principle does apply because they are "serious or irreversible", but then would not demand anything, because they do not threaten to lead to *environmental* damage.

And while the Rio PP talks about threats in general, the Wingspread only refers to *activities* that raise threats, thereby arguably excluding non-anthropogenic threats like, e.g., asteroids. The latter does not have to be a problem, but one has to decide whether this restriction can be defended, i.e., whether it fits with our commitments.

Lastly, the Wingspread PP is unclear with respect to the **knowledge condition**: it states that we should take measures "even if" there is uncertainty, but this could mean that it also applies when we are certain—although taking measures then would arguably no longer count as *precautionary*.

To sum up, pressing questions to be answered are:

- What kinds of threats demand precautions?
- What does count as precautionary measures?
- How should we deal with trade-offs, e.g., if a threat also provides chances of great benefits, or if a precautionary measure introduces new threats?
- When exactly should a PP apply and prescribe measures?

In the next step, B_1, I systematically search for answers to these questions by broadening my set of commitments.

6.4 Step B_1: Adjusting Commitments by Broadening the Current Set

Since there is no current system, commitments cannot be adjusted with respect to account. Instead, the guiding questions that were formulated at the end of step A_1 are used to systematically search for further relevant commitments. This is structured according to the elements of "threat" (Sect. 6.4.1), "knowledge" (Sect. 6.4.2) and "precautionary measures" (Sect. 6.4.3).

While the literature on PPs disagrees on many points, it is commonly accepted that the general structure of a PP includes the three elements of threat, knowledge, and precautionary response (Gardiner 2006; Manson 2002; Randall 2011; Sandin 1999; Steel and Yu 2019). The tripartite PP-structure is part of my initial commitments:

IC 8 The structure of a PP includes two "trigger conditions", threat and knowledge, and a precautionary response. [low]

The two elements of threat and knowledge are also often called "trigger conditions", i.e., that if those two conditions are fulfilled, then precautionary measures are "triggered". Consequently, a lot depends on how those elements are specified. This is closely connected to the last guiding question, which asks *when* exactly a PP should apply and prescribe measures.

The present step, B_1, is structured with respect to the three elements of "threat", "knowledge", and "precautionary response", in order to find some tentative answers (in the form of emerging commitments) to the guiding questions identified in the last step, A_1.

6.4.1 The Element of "Threat"

The notion of "threat" is defined as a "possibility of harm that is uncertain". By examining the threats mentioned in the initial input commitments, further input commitments emerge: the target PP should not be restricted to threats to specific entities, but all serious threats *pro tanto* warrant precaution.

In order to explore which threats are identified as relevant in the commitments, it is first necessary to clarify what a "threat" is. Based on the literature, I adopt Randall's (2011, 31) "chance of harm" concept as a candidate for the conception of *threat*, where harm has the meaning of "damage, impairment" whereas chance "concerns possibilities that are indeterminate, unpredictable, and (in some renditions) unintended". I take that to mean that *threat* encompasses all possibilities of harm that are not certain. In this definition, the uncertainty of the harm neither restricts "threats" to cases where probabilities are available, nor does it exclude them. I choose this conception also because it seems to fit with how *threat* is used in the commitments, but this will have to be assessed when relating the candidate systems to the commitments. Notably, this definition of "threat" is not a commitment, but an attempt at systematizing the subject matter—that is, a part of the candidate system.

Definition 1: Threat A threat is a chance of harm in the sense that there is an indication of possible harm, or a signal correlated with contingent future harm. (Randall 2011, 31–36)

One important open question is what kinds of threats warrant precautions. In my initial commitments (see Appendix A), I refer to threats such as:

- dangerous climate change
- unintended side-effects, both foreseen and unforeseen, of solar radiation management (SRM)
- distributive and intergenerational injustices from SRM implementation
- being shot
- dying in a plane accident
- increased likelihood of getting cancer
- lung cancer
- a small negative impact on the brain development of children
- deleterious effects on organisms in a lake
- unforeseen consequences of a new technology, such as the spread of highly toxic algae
- serious or irreversible damage
- harm to human health or the environment.

Most of these threats actually concern the environment and/or human health, so the question is whether it would be a useful systematization to restrict the target PP to threats to those entities. I argue that no, it is not, at least not for the first attempts: firstly, "distributive and intergenerational injustices" only indirectly concerns aspects of human health; rather the main point here is the injustice. Secondly, it also makes sense to take precautions against, e.g., financial loss. Restricting the target PP only to environmental harm or harm to human health could thus unnecessarily restrict its scope, and a broad scope is one of the desiderata for the system. However, we could wonder whether harm to the environment and/or human health should take lexical priority over other kinds of harms, e.g., that if we have to decide between an action that carries a threat to the environment and an action that carries a threat of economic loss, we should always decide in favor of the environment. But this also seems unduly restrictive, since it could lead to disproportionally huge economic losses for the sake of preventing a negligibly small harm to the environment. Thus, here is an emerging commitment that at the same time might serve to systematize other commitments:

EC 1 The target PP is neither restricted to threats to specific entities (e.g., the environment and/or human health), nor is there a category of threat that takes lexical priority for the application of a PP insofar as it is a threat to specific entities. [low] [emerged at Step B_1]

But what might threats then have in common that makes them warrant precautions? It is important to note that not all of the threats mentioned in the list above are taken to warrant precautions: I am committed to the claims that you should take the job in Chicago even though it means that there is a small probability that you will die in a plane accident, or that radiation therapy for cancer patients is permissible even though it increases the likelihood for them to get cancer again at a later point. This suggests that whether a threat warrants precaution in the sense that the target PP should demand measures be taken cannot solely depend on the severity of the harm, but will also depend on the available evidence as well as on the trade-offs involved.

Still, it should be possible to give some characterization of what kinds of threats *pro tanto* warrant precautions. A *pro tanto* ought is a nonfinal ought that only results in a final ought if there either are no other relevant *pro tanto* oughts or no other relevant considerations, or if it outweighs these other oughts and considerations (Reisner 2013).

Thus, I take it that there are threats that *pro tanto* warrant taking special precautions; but then, based on other information and considerations, this could result in a demand for very minimal measures, or could even be overruled. On that interpretation, threats like dying in a plane accident or getting cancer are of course harms we should take precautions against—it is just that in these specific cases, other considerations overrule the need to avoid those threats.

For now, I am looking for a minimal, qualitative characterization of the kinds of threats that *pro tanto* warrant precaution. A qualitative characterization in terms of so-called "thick" or "value-laden" concepts makes sense, because they highlight that which threats warrant precaution will depend partly on our values, but also on some descriptive characteristics. Although they still require interpretation and deliberation, they facilitate discussion and provide focus (Gardiner 2006, 57–58).

Candidates for such a characterization of threats are, e.g., serious harm, irreversible harm, unacceptable outcomes, or catastrophic outcomes.

As a commitment, I accept that *serious* threats *pro tanto* warrant precautions. This qualitative characterization seems useful to me, since a threat is—according to Definition 1—a possibility of harm that is *uncertain*. Since this entails that we are not sure whether or not harm will occur, it makes sense to focus on threats that are in some way *serious*, i.e., cases in which the costs of being wrong are significant.

EC 2 Serious threats *pro tanto* warrant precaution. [low] [emerged at Step B_1]

While threats of unacceptable or catastrophic outcomes are arguably also serious threats, irreversible harm is not always serious and is sometimes even completely negligible. Although irreversibility of harm can make a threat more serious, it is not plausible that it should in itself warrant precautions.[2]

For the understanding of "serious", I commit to the proposal of Resnik (2003): seriousness is assessed according to (i) the potential for harm of the threat, and (ii) whether or not the potential damage is seen as reversible. This also allows us to compare threats.

EC 3 The seriousness of a threat is assessed according to (i) the potential for harm of the threat, and (ii) whether or not the possible harm is seen as reversible. [low] [emerged at Step B_1]

Perhaps more characteristics can be named that make a threat serious, e.g., how its potential for harm can be assessed, but this seems like a good first formulation.

It will then be a task for the target precautionary principle to identify against which threats that *pro tanto* warrant precautions we actually should take precautions, and to what extent.

The next important step towards this is to address the level of knowledge—respectively uncertainty—at which the target PP should demand measures.

6.4.2 The Element of "Knowledge"

> As a minimal knowledge level, i.e., what we have to know about a threat in order for that threat to warrant precaution, *plausibility* is selected. For a threat to be assessed as plausible, we need at least some credible scientific evidence in its favor, even though it might not be enough to assign probabilities.

One of the core ideas of PPs is that we do not have to—and often should not—wait for full scientific certainty before taking measures to prevent harm. Consequently,

---

[2] See Randall (2011, 57–72) for a discussion of the notion of irreversibility.

the level of knowledge required to "trigger" precautionary measures should be something less than full (scientific) certainty. On the other hand, it should be more than mere logical possibility: if it is only required that a threat is logically possible, then virtually every action or inaction can lead to catastrophic harm—my writing of this book might by some ludicrous, but logically possible, chain of events lead to a nuclear holocaust.

One popular approach to settling the knowledge condition for a precautionary principle is to argue that it applies under conditions of decision-theoretic uncertainty, meaning decision situations in which we have knowledge of the available courses of actions along with their complete set of possible outcomes, but cannot assign probabilities to those outcomes. This is often combined with suggesting a "division of labor" with quantitative approaches like cost-benefit analysis, that can be applied in situations of decision-theoretic risk, where we also have knowledge of the probabilities of the possible outcomes. However, neither the risk/uncertainty distinction nor the use of decision-theoretic uncertainty as the knowledge condition for PPs is without critics (Randall 2011; Roser 2017; Steel 2015); and it has been argued that there are situations where a PP should apply even though probabilities are available (e.g., Randall 2011; Steel 2015; Thalos 2012). For now, I do not want to take a stance on this by committing to whether or not the risk/uncertainty distinction is relevant for the application set of the target PP. This is rather a question that should be addressed during the course of this process.

However, it makes sense to set a minimal knowledge level at which serious threats *pro tanto* warrant precautions. This lower boundary should be in some meaningful sense more than mere logical possibility without yet presupposing that relative likelihoods or probabilities are available. For this purpose, the notion of *plausibility*, or credibility, of a threat seems suitable. For a threat to be assessed as plausible, we need at least some credible scientific evidence in its favor, even though it might not be enough to assign probabilities. Consequently, the *plausibility* of a threat is not to be confused with its *likelihood* (Resnik 2003, 340–41). While a plausible serious threat *pro tanto* warrants precaution, how extensive the measures are that we take then might just as well depend, *inter alia*, on how likely we judge the threat to be. But I argue that the target PP should *pro tanto* apply to all relevant *plausible* threats, and adopt the following commitment:

EC 4 All *plausible* serious threats *pro tanto* warrant precaution. [low] [emerged at Step B_1]

Of course, what "plausible" means needs to be spelled out more—it could refer to, e.g., epistemic and pragmatic criteria to assess the plausibility of a hypothesis (Resnik 2003), or that we have to know a mechanism by which the threat would be realized (Hartzell-Nichols 2017), or that we cannot show that the threat is inconsistent with our scientific background knowledge (Betz 2010). However, explicating the notion of "plausibility" is outside the scope of my current epistemic project, which focuses on formulating a moral precautionary principle. In continuing, then, I thus stipulate that there is a meaningful notion of plausibility in the background, respectively that the explication of such a notion has no direct implications for the formulation of the target PP in the current RE process.

6.4.3 *The Element of "Precautionary Measures"*

The set of input commitments is extended by a range of emerging input commitments concerning what does, or does not, count as a precautionary measure.

I am committed to a range of measures, but almost none of them are explicitly characterized as being precautionary. When trying to find clear-cut cases of precaution, it can be difficult to distinguish precautionary measures—of the kind the target PP should demand—from everyday caution, as well as from taking preventive measures where not taking them would simply be careless, reckless, or outright suicidal.

In order to elicit more commitments that explicitly concern whether or not a measure is precautionary, I did consider cases where people might tell you to be cautious, or might rebuke you, saying that you should have been more cautious if something happens to you because you did not take specific measures or actions—but these measures nonetheless might not count as precautionary measures. You can find the full list of emerging commitments in the Appendix A, but here are some examples:

EC 6 Looking left and right before crossing the street is not a precautionary measure. [low] [emerged at Step B_1]

EC 7 To bring a parachute when planning to jump out of an airplane is not a precautionary measure. (Example from Sandin 2004) [medium] [emerged at Step B_1]

EC 8 To have a parachute on board when planning to fly somewhere is a precautionary measure. (Example from Sandin 2004) [medium] [emerged at Step B_1]

EC 11 Chewing your food is not a precautionary measure against choking. [medium] [emerged at Step B_1]

EC 12 As a factory worker who is well informed about the dangers of being exposed to the hazardous chemical X in your work, performing a ritualistic dance to protect you from a hazardous chemical is not a precautionary measure. (Example from Sandin 2004) [high] [emerged at Step B_1]

I also identified a more general commitment about precautionary measures:

EC 13 Precautionary measures should be effective in preventing or substantially ameliorating either a threat or the harm of a threat. [high] [emerged at Step B_1]

6.4.4 *The Broadened Set of Current Commitments, C_1*

At the end of step B_1, the current set of commitments, C_1, consists of the initial commitments C_0 (none of them having so far been adjusted) and 13 emerging commitments, EC 1–EC 13. Among these, EC 5–EC 13 specifically concern what does or does not count as a precautionary measure. However, we are still lacking clear criteria for what makes a measure a *precautionary* measure.

In the next step, A_2, an explication for "being a precautionary measure against an undesirable *x*" is proposed as a partial systematization of the subject matter which is constrained by the commitments EC 5–EC 13 and further emerging commitments. Steps A_2 and B_2 explicate what the necessary and jointly sufficient conditions are for an action to count as a *precautionary* measure. However, it does not follow that every precautionary measure is *warranted.* Identifying which precautionary measures are justified, respectively required, from a moral standpoint, is then the purpose of the target system as a whole.

6.5 Step A_2: Explicating "Precautionary Measures"

Steps A_2 and B_2 are an explication of the concept of "being a precautionary measure against an undesirable *x*." They are thus a "sub-process", i.e., they concern only a part of the position. Consequently, in this step, A_2, a candidate for a *part* of the system, and not a candidate for the whole system, is suggested. It has a smaller scope and is only supposed to systematize commitments concerning what does or does not count as a "precautionary measure". I use the explication proposed by Sandin (2004), and after assessing it with respect to account (Sect. 6.5.1) and its theoretical virtues (Sect. 6.5.2), it is adopted as part of the system.

Some characteristics that the commitments concerning precautionary measures have in common seem to be: the action has to be performed intentionally (if you bring a fire-extinguisher as part of your costume, this was not a precaution against a sudden fire outbreak at the party, EC 9), there has to be sufficient uncertainty about whether or not the threat would materialize if the measures were not taken (EC 7: Bringing a parachute when you plan to jump out of an airplane is not a precautionary measure), and the measures should in fact eliminate or at least diminish the threat (EC 13: Precautionary measures should be effective in preventing or substantially ameliorating a threat, and EC 12: A ritualistic dance is not a precautionary measure against threats from chemicals). These aspects have been identified as necessary and jointly sufficient conditions for some action to be a precautionary measure against an undesirable *x* by Sandin (2004):

ExplicPrec *Explication of "being a precautionary measure against an undesirable* x*"*:
An action *a* is precautionary with respect to something undesirable *x* if *a* fulfills the following necessary and jointly sufficient criteria:

1. Intentionality: *a* is performed with the intention of preventing *x*.
2. Uncertainty: the agent does not believe it to be certain or highly probable that *x* will occur if *a* is not performed.
3. Reasonableness: the agent has externally good reasons (a) for believing that *x* might occur, (b) for believing that *a* will in fact at least contribute to the prevention of *x*, and (c) for not believing it to be certain or highly probable that *x* will occur if *a* is not performed.

I adopt *ExplicPrec* as a candidate for the explication of precautionary measures, and in the following assess it with respect to account for commitments and its theoretical virtues.

6.5.1 Account for Commitments About Precautionary Measures

ExplicPrec accounts for all of the **emergent commitments** on precautionary measures, i.e., EC 5–EC 13. Take the examples from before:

EC 6, crossing the street: we know that if one does look left and right before crossing the street, it is very likely that at some point one will be hit by other road users. Not looking left and right before crossing the street is reckless. Hence, the fact that it does not count as a precautionary measure fits with the explication: the *uncertainty* criterion is not fulfilled.

EC 7, bringing a parachute when planning to jump out of a plane: we know that people die if they jump out of an airplane without a parachute at 4000 meters. Thus, taking a parachute is not a precaution against uncertain harm: not taking one would be outright suicidal. This fits with the explication: the *uncertainty* criterion is not fulfilled: we believe it to be certain that people die when they jump out of airplanes without a parachute.

EC 8, bringing a parachute when flying somewhere: we know that it is possible that planes have accidents that make it necessary to jump off with a parachute to save oneself, but we do not expect that it will happen this time. Still, we bring the parachute as a precautionary measure against dying in a plane crash. This fits with the explication, all three criteria are fulfilled.

EC 11, chewing your food carefully: we know that people can choke if they don't chew their food properly, but wolf it down. So it again fits with the explication, because the *uncertainty* criterion is not fulfilled.

EC 12, performing a ritualistic dance: at least in our society, there are no good reasons for the factory worker to believe that this will prevent them from harm. Thus, the commitment that it does not count as a precautionary measure against the hazardous chemical is accounted for by the explication, since the *reasonableness* criterion is not fulfilled.

Through the *Intentionality*- and the *Reasonableness* criterion, *ExplicPrec* also accounts for the commitment EC 13, i.e., that precautionary measures should be effective in preventing or substantially ameliorating either a harm, or the threat of a harm.

In the **initial set of commitments**, there were a few commitments that included some relevant aspects of precautionary measures and that need to be considered here as well. There weren't any real conflicts, even if the explication could not always straightforwardly account for them. But the latter is rather caused by the fact that we do not yet have a full system that could account for all aspects of the commitments. The relevant commitments are IC 23 and IC 24:

IC 23 Independently of whether or not SRM should be considered as part of precautionary measures in case the globally implemented mitigation and adaptation strategies turn out to be insufficient to prevent dangerous climate change, it should not be used as the only precautionary measure. ("SRM" here is short for "research and development on solar radiation management in order to have it ready to use should dangerous climate change be imminent".) [high]

This commitment expresses that SRM on its own is not enough as a precautionary measure, but it does not tell us whether or not it *is* a precautionary measure. To determine this, we can now use the explication. However, assessing the implications of the current system for the current set of commitments will be discussed in step B_2.

IC 24 SRM should not be considered as the only precautionary measure against the threat that the globally implemented mitigation and adaptation strategies turn out to be insufficient to prevent dangerous climate change, because it is inadequate as a precautionary measure. It is inadequate because: it introduces threats of its own, it is uncertain whether it would work in the intended way without unforeseen (negative) side-effects, and it imposes costs and responsibilities (e.g., for maintenance) on future generations. [medium]

This commitment claims again that SRM on its own is not a precautionary measure against dangerous climate change, and it basically states that aspect (b) of the reasonableness criterion is not fulfilled: it is uncertain whether SRM would work in the intended way, i.e., the reasons to believe that it will in fact prevent dangerous climate change are not good enough. However, the explication cannot account for the claim that SRM is inadequate because it imposes costs and responsibilities on future generations. This might be a sign that there is something that makes SRM inadequate as a measure not from a precautionary perspective, but on some other grounds.

6.5.2 *Theoretical Virtues of the Precautionary-Measures Explication*

The theoretical virtues of *ExplicPrec* have to be assessed with respect to it being part of the overall target system. Since there is no alternative candidate for the concept of "precautionary measures", and not yet a current system that the explication can be assessed as a part of, I assess the virtues of the explication on its own, in order to decide whether there is something that speaks strongly against it.

Determinacy There can be boundary cases about what does count as "certain or highly probable" (the *Uncertainty* condition), in part because this might be context-dependent. The same holds for "externally good reasons" (the *Reasonableness* condition). But aside from this, the criteria seem clear-cut and precise—at least clear-cut and precise enough to contribute to the pragmatic-epistemic objective.

Practicability That the two criteria of *Intentionality* and *Uncertainty* refer to the intentions and beliefs of the agent could cause a problem for practicability, in the sense that we do not have direct access to the intentions and beliefs of others. But as long as we understand *Intentionality* as saying that nothing can be a precautionary measure against *x* as long as it has not at least been declared to be intended to prevent *x*, this is not a real problem. And the *Uncertainty* criterion also has to be checked by the *Reasonableness* criterion. I.e., even for other agents we can assess whether or not they can reasonably see a measure they intend to take as a precautionary measure.

Broad Scope The explication has a broad range of applicability, since its criteria 1.–4. are necessary and jointly sufficient: for every combination of fulfilled or not fulfilled criteria, it will tell us whether a measure is precautionary against a specific undesirable *x*, or not.

Simplicity I take it that the explication includes the following concepts that are not reducible to each other: intention; preventing (an event); belief; certain or highly probable; externally good reasons (for believing something). While not being extremely minimal, this is not a level of complexity that would be high enough to diminish any of the other virtues, as their assessment has shown.

Evaluating *ExplicPrec* Since there was no other candidate, and the explication of precautionary measures against an undesirable *x* did do very well with respect to accounting for relevant commitments, and reasonably well with respect to theoretical virtues, I argue that it should be adopted as (a part of) the system, i.e., as the current partial system S_1.

6.6 Step B_2: Adjusting Commitments About Precautionary Measures

The subset of the current commitments that concern precautionary measures are adjusted with respect to the newly chosen explication of "being a precautionary measure against an undesirable *x*". By making a further emerging commitment explicit, the account value can be increased (Sect. 6.6.1). Additionally, the explication has more implications than are part of the current commitments. Those implications concerning current commitments are added as newly inferred commitments (NCs) (Sect. 6.6.2). Lastly, the set of emerging input commitments is further broadened by exploring my commitments concerning which precautionary measures are *warranted* (Sect. 6.6.2).

In this step, the current set of commitments C_1 is adjusted with respect to the current partial system S_1, the explication of "precautionary measures", *ExplicPrec*.

First, are there ways to adjust the current set of commitments in order to increase agreement with the current partial system?

6.6.1 *Trying to Increase Account*

Above, I stated that the explication cannot directly account for IC 23:

IC 23 Independently of whether or not SRM should be considered as part of precautionary measures in case the globally implemented mitigation and adaptation strategies turn out to be insufficient to prevent dangerous climate change, it should not be used as the only precautionary measure. ("SRM" here is short for "research and development on solar radiation management in order to have it ready to use should dangerous climate change be imminent".) [high]

This commitment only directly entails that SRM on its own is not enough as a precautionary measure, but not whether or not it should be considered as a precautionary measure (just maybe an inadequate one). But when thinking about this, it becomes clear to me that I am also committed to the following:

EC 14 Research and development (R&D) into solar radiation management (SRM) in order to have it ready to use should dangerous climate change be imminent, is not, on its own, a precautionary measure against the threat of dangerous climate change. [medium] [emerged at Step B_2]

This means that there is a further emerging commitment that needs to be added to the set of current commitments (and the set of input commitments). Can the explication account for EC 14? The *Intentionality* and the *Uncertainty* criterion are fulfilled, but what about the *Reasonableness* criterion? Do I have (a) externally good reasons for believing that dangerous climate change might occur?—According to my background information, yes. Do I have (b) good reasons for believing that R&D into SRM will in fact at least contribute to the prevention of dangerous climate change?—This depends on what "good reasons for believing" means, because there are reasons for believing that SRM can substantially counteract global warming caused by alleviated GHG levels. But there are also reasons to be concerned that it might have unforeseen, unintended negative side-effects that would prevent it from working in the intended way, maybe even such that it adds to the harmful impacts of dangerous climate change. Since there are (c) externally good reasons for not believing it to be certain or highly probable that dangerous climate change will happen if R&D into SRM is not performed, whether or not SRM counts as a precautionary measure thus depends on how we evaluate aspect (b) of the *Reasonableness* criterion. I stipulate that according to my background knowledge, the uncertainty about the effectiveness and the potential side-effects of SRM and its R&D is just too high to fulfill this criterion. Consequently, SRM on its own is not a precautionary measure against the threat of dangerous climate change—the explication can account for this commitment.

6.6.2 Newly Inferred Commitments that Classify Measures as (Not) Precautionary

In the case of most of my commitments I wasn't yet committed on whether or not the endorsed or rejected measures counted as precautionary. Adopting the explication of "being a precautionary measure against an undesirable *x*" thus leads to a range of newly inferred statements that I would have to accept as newly inferred commitments if I were to adopt the explication as a part of the system. This actually generates a lot of new commitments, since every measure that I endorse or reject in my current commitments is classified as precautionary or not precautionary by the explication. Here are some examples (for a full list, see Appendix A):

NC 2 Requiring that any application of SRM against harmful impacts of climate change has to be accompanied by a strict mitigation and adaptation program (IC 6) is not a precautionary measure against other effects of increased GHG levels and negative effects from prolonged SRM-implementation. (Uncertainty about those negative effects is too low.)

NC 9 In case 3, *R&D into SRM, two kinds of research*, choosing option (iii), not implementing any research and/or development program into SRM, does count as a precautionary measure against the potential dangers of a full-blown R&D program into SRM.

NC 10 In case 3, *R&D into SRM, two kinds of research*, choosing option (iii), not implementing any research and/or development program into SRM, does count as a precautionary measure against the possibility that a non-invasive research program into SRM turns out to be a waste of money and effort that does not help us to prevent dangerous climate change.

NC 13 In case 1, *Genetically Engineered Algae*, banning the technology is a precautionary measure against possible harmful effects from it.

NC 14 In case 9, *Job Offers*, choosing the job in Chicago is not a precautionary measure against having a tedious and badly paid job in New York. (By design of the case, it is certain that you end up with the bad job if you don't go to Chicago.)

NC 15 In case 9, *Job Offers*, choosing the job in New York is a precautionary measure against being killed in a plane accident.

These "newly inferred commitments" are commitments that I accept because they follow from my current system. But they are not input commitments—independently of the current system, I would not have come up with them.

6.6.3 Searching for Further Relevant Commitments

ExplicPrec tells us what a precautionary measure is, but not whether or not a precautionary measure is *warranted*. This is something that the target system as a whole will have to address. For this purpose, we can already search for further relevant commitments.

First, I am committed to the claim that precautionary measures should not introduce serious threats of their own. I do, however, only assign a low weight to this commitment: I think it is a possibility that there might be further constraints that can make it defensible that a precautionary measure introduces serious threats of its own—depending on the trade-offs involved.

EC 15 Precautionary measures should not introduce serious threats of their own. [low] [emerged at Step B_2]

Also, I am committed to the claim that in order to be defensible, costs and responsibilities for precautionary measures should be distributed in a morally sound way, e.g., it should not be the case that the general public has to pay for precautionary measures against a threat caused by an action that will only benefit a very small minority.

EC 16 The costs and responsibilities for precautionary measures should be distributed in a morally sound way. [high] [emerged at Step B_2]

When talking about additional threats, and the costs of precautionary measures, it seems that there is a *price* that we have to pay for a precautionary measure, and the objective of a PP is also to tell us when this price is adequate (cf. Munthe 2011).

EC 17 The price of a precautionary measure consists of—compared with the course of action entailing the threat it is supposed to address—foregone benefits,[3] foregone opportunities, and additional threats. [medium] [emerged at Step B_2]

And this price should be proportional to what is at stake:

EC 19 The price of precaution should be proportional to the seriousness and the plausibility of the threat, given the available alternatives. [low] [emerged at Step B_2]

Based on this discussion, I am now also making explicit the following general commitment concerning the target system:

EC 20 The target PP applies to plausible and serious threats and prescribes measures that are proportional to the severity and plausibility of the threat. [medium] [emerged at Step B_2]

6.6.4 The Adjusted Set of Current Commitments, C_2

At the end of step B_2, the current set of commitments, C_2, consists of all the commitments that were in C_1, thirty newly inferred commitments on what does and does not count as a precautionary measure against an undesirable x, and six more emerging commitments concerning justified precaution and demands on the target PP, EC 15–EC 20.

Thus, the current commitments C_2 consist of the following subsets:

1. all the initial commitments of C_0, i.e., IC 1–31, *plus*
2. ten emerging commitments specifying constraints and desiderata for the target system (EC 1–EC 4; EC 15–EC 20),
3. ten emerging commitments on what does and does not count as a precautionary measure against some x (EC 5–EC 14), and
4. thirty newly inferred commitments on what does and does not count as a precautionary measure against some x (NC 1–NC 30).

Currently, we were only adjusting commitments with respect to a part of the system, namely the explication of "being a precautionary measure against an undesirable x". The relevant subsets of the current commitments are 3 and 4, and both these subsets are accounted for by the explication.

[3] I take it that "foregone benefits" also includes direct monetary costs of precautionary measures that are spent, e.g., on installing safety measures, since the money used there cannot be spent for other purposes.

6.7 Step $A_{3.1}$: Formulating the Principle 3 Candidate System

A new candidate system, the Principle 3-System, is formulated by combining the explication of "precautionary measures" with results from answering the guiding questions.

Step A_3 is "split" between phases 1 and 2, since it on the one hand includes the formulation of a candidate system based on steps A_1–B_2, which still belongs to phase 1. I call this part step $A_{3.1}$. On the other hand, this candidate system will then be compared with further candidates that are taken from the literature. While still being a part of step A_3, the latter introduces the beginning of my phase 2—thus being labeled step $A_{3.2}$.

At the end of step A_1, the following open issues for the formulation of a candidate system were identified:

- What kinds of threats demand precautions?
- What does count as precautionary measures?
- How should we deal with trade-offs, e.g., if a threat also provides chances of great benefits, or if a precautionary measure introduces new threats?
- When exactly should a PP apply and prescribe measures?

Based on the results from steps B_1–B_2, I suggest the following candidate principle:

Principle 3 (P 3) Where there are plausible threats of serious harm, precautionary measures that are proportional to the severity and plausibility of the threat should be taken.

Arguably, P 3 has more determinacy than P 1 and P 2 because it at least somewhat specifies the precautionary measures that should be taken. By demanding that measures should be *proportional* it also somewhat addressed the problem of threat trade-offs, i.e., taking precautionary measures that themselves threaten to cause more harm than the original threat.

Of course, I also have to say a bit more about the concepts used in P 3, or it won't be much more determinate than P 1 or P 2. In order to do that, I add some further parts to the system.

Here I can draw on the results from B_1, when the set of commitments was systematically broadened with respect to the guiding questions. As a candidate for the conception of *threat*, I adopt my commitment to Randall's "chance of harm" concept, where harm has the meaning of "damage, impairment", whereas chance "concerns possibilities that are indeterminate, unpredictable, and (in some renditions) unintended" (Randall 2011, 31):

P 3.1: Definition: Threat A threat is a possibility of harm that is uncertain.

For the understanding of *serious*, I follow Resnik (2003): seriousness is assessed (i) according to the potential for harm of the threat, and (ii) whether or not the potential damage is seen as reversible. This also allows the comparison of threats.

P 3.2: Seriousness of Threats The seriousness of a threat is assessed according to (i) the potential for harm of the threat, and (ii) whether or not the possible harm is seen as reversible. [same content as IC 11]

For a threat to be assessed as *plausible*, we need at least some credible scientific evidence in favor of this, even though it might not be enough to assign probabilities. Consequently, the *plausibility* of a threat is not to be confused with its *likelihood* (Resnik 2003, 340–41). While a plausible serious threat *pro tanto* warrants precaution, how extensive the measures are that we take then might just as well depend, *inter alia*, on how likely we judge the threat to be. But I argue that the target PP should *pro tanto* apply to all relevant *plausible* threats. Together with *ExplicPrec*, we then have a new candidate system, which I call the Principle 3-System.

P 3.3: *ExplicPrec* Explication of "Being a precautionary measure against an undesirable *x*": An action *a* is precautionary with respect to something undesirable *x* if *a* fulfills the following necessary and jointly sufficient criteria:

1. Intentionality: *a* is performed with the intention of preventing *x*.
2. Uncertainty: the agent does not believe it to be certain or highly probable that *x* will occur if *a* is not performed.
3. Reasonableness: the agent has externally good reasons (a) for believing that *x* might occur, (b) for believing that *a* will in fact at least contribute to the prevention of *x*, and (c) for not believing it to be certain or highly probable that *x* will occur if *a* is not performed.

The Principle 3-System thus consists of the four parts of P 3, P 3.1, P 3.2, and P 3.3.

6.8 Recapitulation Phase 1

Formulating the Principle 3-System concludes the first phase of my RE application, in which I explore how candidate systems can be constructed in the RE framework. The results of the steps from phase 1 are summarized in the schematic overview of Fig. 6.1, starting from the initial set of commitments C_0 to the formulation of the P 3-System at the beginning of step A_3. In the following, I discuss what can be learned from this both for RE and PPs, before moving to phase 2 in Chap. 7.

I first discuss what the intermediate results for RE and its application are (Sect. 6.8.1)—my goal *with* its application, i.e., the case study—before discussing some results for (moral) precaution and precautionary principles (Sect. 6.8.2)—my goal *in* the RE application.

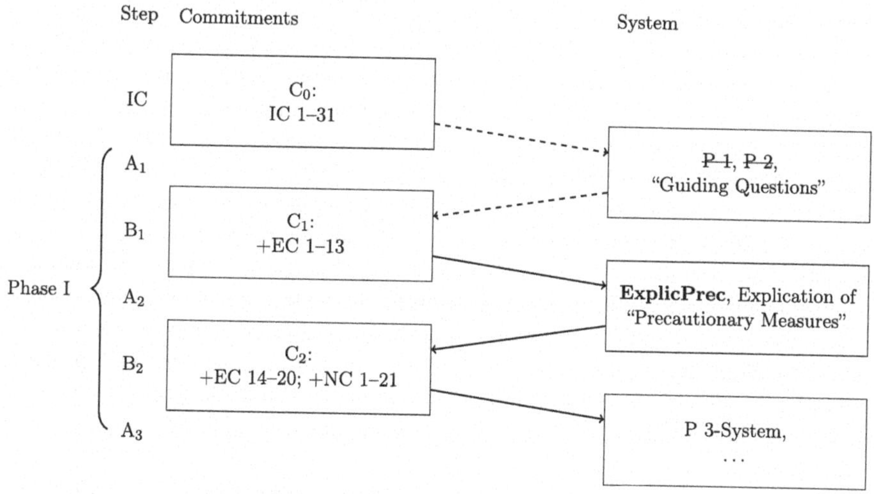

Fig. 6.1 Schematic overview of the steps of Phase 1

6.8.1 Phase 1: Discussion of Intermediate Results for RE

The following intermediate results for RE are discussed:

- Background information and background theories can play important roles for, e.g., how candidate systems are interpreted and assessed;
- emerging input commitments played a central role: systematically broadening the set of commitments as part of the RE process;
- the difference between commitments and (parts of) the system: a difference of function, not form or content;
- the construction and development of systems can be made part of an RE process.

Firstly, one result of assessing the Rio and the Wingspread PP formulations as Principle 1 and Principle 2 is that it demonstrates the **relevance of background information** and of background **theories**: If we could presuppose a lot more, then Rio or Wingspread would be more determinate and reach a higher account value. Indeed, as regulatory principles, they, and other similar formulations, stand in a specific context and (legal) practice, which might often make them more determinate than assessing them as stand-alone (moral) principles suggests (cf. Fisher 2002). However, the pragmatic-epistemic objective of this RE implementation is not to justify a principle for regulatory policy, but an action-guiding *moral* principle.

Secondly, some results with respect to **commitments**: Emerging commitments played an important role—a factor that is typically neglected in RE conceptions (see Chaps. 1 and 2). Ideally, an RE process would be conducted with respect to all commitments that the epistemic agent holds. But since this is impossible for several reasons (e.g., problems with individuating commitments, comprehensibility and manageability of the process, cognitive limitations, implicit commitments, etc.), it can only be done with respect to the commitments that are explicitly considered. Even if those are carefully selected with respect to being, e.g., as representative as possible, it is still likely that relevant commitments are missing. Thus, in the first two B-steps of the RE implementation, B_1 and B_2, adjustments to the set of current commitments did consist in broadening the set by making further input commitments explicit and adopting commitments as inferences from the system. One could object that this is not really a part of RE, but rather a part of selecting the initial commitments. In this view, steps A_1–B_2 aren't already part of an RE process, but rather of identifying relevant initial commitments to start the process. EC 1–20 then would not count as emerging commitments, but should simply be part of the initial commitments.

However, I argue that since we cannot make everything explicit from the beginning, we have to start *somewhere*, and making further commitments explicit is necessarily part of applying RE to actual, complex problems. That further commitments concerning what does or does not count as a precautionary measure would be relevant only became apparent when assessing the Rio and the Wingspread PP as first candidate systems in the form of Principle 1 and Principle 2. This also shows that starting with other candidate principles might have led to other emerging commitments, putting the process on another pathway.

A further insight concerns the **difference between commitments and (partial) systematizations**. The commitments that emerge with respect to the guiding questions from step A_1 support that the difference between commitment and system is a difference of function, but not of form or content: most of them are candidates for systematizing other commitments, while at the same time I am committed to them too. Take for example EC 4:

EC 4 All *plausible* serious threats *pro tanto* warrant precaution. [low] [emerged at Step B_1]

While expressing my (low) commitment to the claim that all *plausible* serious threats *pro tanto* warrant precaution, EC 4 also can serve as a candidate for a partial systematization of other commitments, e.g., to distinguish between threats that *pro tanto* warrant precaution and those that do not. Depending on which function it is supposed to fulfill, different constraints apply to it: as a commitment, I have to respect its independent credibility when trying to bring it into agreement with a system. As a candidate for a part of the system, it has to prove successful in systematizing commitments by accounting for the relevant subsets of them, while also having theoretical virtues (respectively contributing to the theoretical virtuousness of the system as a whole). In its role as a candidate for a part of the system, it does, however, not have independent credibility and can always be given up.

Thirdly, the results from phase 1 show how the **construction and development of candidate systems** can be part of an RE process: normally, in RE conceptions, it is presupposed that we compare and adjust principles, but it is not described how we obtain them. As a creative element, the proposal of candidate systems is not seen as a part of the RE process. It might indeed be true that formulating candidate systems is a creative process that cannot be rule-governed and thus cannot be described in explicit terms as part of the RE steps. However, the steps of phase 1 show how the RE criteria can provide heuristics and guidelines for the development of candidate systems: they might, for example, show us how commitments can be systematically explored with respect to questions and problems that available candidate systems leave open, and how systems can be step-by-step constructed by sub-processes like explications. The formulation of **guiding questions** is not something that is usually described as part of RE, but it proved to be helpful for the formulation of candidate principles. In a sense, these questions themselves are (very preliminary) candidate systematizations, since they suggest what the relevant factors might be that one should explore further.

The first phase also demonstrates how the holistic process of justification via RE often has to proceed piecemeal, since we often have to clarify one specific aspect before being able to move on with the bigger picture. It thereby also shows how seemingly piecemeal and isolated work can be part of a more holistic process.

6.8.2 *Phase 1: Discussion of Intermediate Results for Precaution*

> The main intermediate result for the discussion of precautionary principles is that measures are not automatically justified qua being precautionary.

As the main result for precaution from phase 1, we can note that **being "precautionary" does not necessarily amount to being justified**: what is striking when assessing the agreement between the explication and the current commitments is that the explication does not classify every measure that I already endorse as precautionary, and actually does classify some rejected measures as precautionary. However, I do not see this as a problem.

First of all, I see it as an advantage that when accepting the explication of "Being a precautionary measure against an undesirable *x*", measures are not automatically justified just qua being precautionary. There are cases of unwarranted precaution, and it will be part of the task for the target precautionary principle to distinguish them from the warranted ones.

Secondly, there are two reasons why measures expressed in my commitments are classified as not being precautionary: either (1) because they do not fulfill the

Reasonableness criterion, e.g., because there are no good reasons for believing that those measures would in fact contribute to the prevention of the threat they are aimed at. Or (2) because the *Uncertainty* criterion is not fulfilled and it is actually highly probable or even certain that the threat would materialize if the measures were not taken.

With respect to (1), this is in line with my commitments, i.e., the measures that are classified as not being precautionary on grounds that they are not reasonable in some respect are also rejected in the commitments. With respect to (2), these are measures that seem to be demanded on some other, maybe stronger grounds than precaution: while the first group is not precautionary because it seems questionable whether it would avoid the threat, the second group is not precautionary because it is clear that without them the threat would materialize. That is why those measures are not precautionary, even though not doing them would be completely negligent, respectively would mean to accept the negative consequences knowingly.

The question is now how to handle these commitments that endorse the second kind of non-precautionary measures. Do they still belong to the subject matter in the sense that the target system has to account for them, and has to account for them in the sense that they should be inferable from the system?

It seems clear that the target PP in any case should not recommend the opposite of those measures. In that sense they are like control stations—if the target PP recommends against them, then it is plausible that something is very, very wrong. Nonetheless, it does not necessarily have to recommend them: the most important part is that it covers the cases in which *precautionary* measures are warranted, and is able to distinguish warranted from unwarranted precaution. However, since a broad scope is a desideratum for the system, if it can cover all the relevant cases of precaution *and* further cases, then all the better.

All in all, I argue that accepting the explication of "being a precautionary measure against an undesirable *x*", *ExplicPrec*, provides us with a first important systematization of part of the subject matter, picking out a class of relevant cases from related, similar ones. However, it would of course be desirable to find a system that can pick out justified cases of precautionary measures *without* having to refer to such an additional explication: for reasons of simplicity and maybe also practicability, having fewer additional parts in the system is desirable.

References

Ahteensuu M (2008) The precautionary principle and the risks of modern agri-biotechnology. In: Launis V, Raikkä J (eds) Genetic Democracy. Philosophical Perspectives. Springer, Dordrecht, pp 75–92

Betz G (2010) What's the worst case? The methodology of possibilistic prediction. Analyse und Kritik 32(1):87–106

Fisher E (2002) Precaution, precaution everywhere: developing a common understanding of the Precautionary principle in the European Community. Maastricht Journal of European and Comparative Law 9(1):7–28

Gardiner SM (2006) A core Precautionary principle. J Polit Philos 14(1):33–60
Hansson SO (2016) Evaluating the uncertainties. In: Hansson SO, Hirsch Hadorn G (eds) The argumentative turn in policy analysis. Springer, Berlin, pp 79–104
Hartzell-Nichols L (2017) A climate of risk: Precautionary principles, catastrophes, and climate change. Routledge, New York
Manson NA (2002) Formulating the precautionary principle. Environ Ethics 24(3):263–274
Munthe C (2011) The price of precaution and the ethics of risk, the international library of ethics. In: Law and Technology, vol 6. Springer, Berlin
Randall A (2011) Risk and Precaution. Cambridge University Press, New York
Reisner AE (2013) Prima Facie and Pro Tanto Oughts. In: Lafollette H (ed) The international encyclopedia of ethics. Blackwell Publishing Ltd, Hoboken, pp 4082–4086. https://doi.org/10.1002/9781444367072.wbiee406
Resnik DB (2003) Is the precautionary principle unscientific? Studies in History and Philosophy of Science Part C: Studies in History and Philosophy of Biological and Biomedical Sciences 34(2):329–344
Roser D (2017) The irrelevance of the risk-uncertainty distinction. Sci. Eng. Ethics 23(5):1387–1407. https://doi.org/10.1007/s11948-017-9919-x
Sandin P (1999) Dimensions of the precautionary principle. Hum. Ecol. Risk Assess. Int. J. 5(5):889–907
Sandin P (2004) The precautionary principle and the concept of precaution. Environmental Values 13(4):461–475
Sandin P, Peterson M, Hansson SO, Rudén C, Juthe A (2002) Five charges against the precautionary principle. J. Risk Res. 5(4):287–299
Steel D (2015) Philosophy and the Precautionary principle. Cambridge University Press, Cambridge
Steel D, Yu J (2019) The Precautionary principle meets the hill criteria of causation. Ethics, Policy and Environment 22(1):72–89. https://doi.org/10/ggkp6k
Thalos M (2012) Precaution has its reasons. In: Kabasenche W, O'Rourke M, Slater M (eds) Topics in contemporary philosophy 9: the environment, philosophy, science and ethics. MIT Press, Cambridge, pp 171–184

Chapter 7
Case Study, Phase II: Focus on the Process of Adjustments

The previous Chap. 6, explored how the RE steps can be used for the development of a candidate system, i.e., theory *construction*. The present chapter is the second phase of the case study. It focuses on how the RE criteria can be used to comparatively assess different candidate systems, i.e., theory *choice*, and to adjust commitments and system with respect to each other. In the next Chap. 8, a specific position will be fleshed out, and evaluated with respect to whether or not it is a justified position in RE.

7.1 Overview: Phase 2

In the second phase, I focus on implementing the two alternating steps of adjusting the system with respect to commitments (step A) and vice versa (step B) while being as detailed as possible. In order to have enough input in form of candidate systems to compare, I adopt proposals from the PP literature (see Chap. 4). Considering real alternatives, i.e., compelling rival principles, is important to provide a fuller justification in RE (cf. Knight 2017, 50). In the context of the case study, this is also important in order to test whether the RE criteria can be applied in a meaningful way to make a selection between different strong candidates. Additionally, I could have tried to obtain further candidates in ways similar to how the first candidate was developed in phase 1, or through exploring different ways for, e.g., increasing the theoretical virtues of the Principle 3-System. Developing a sophisticated candidate system takes time, so adopting candidates from the literature is not only a way to make sure that real alternatives are considered in order to avoid dealing with straw men; it is also a pragmatic decision in the context of this being a case study for RE—developing a broad range of original candidates would take up a lot of time and space.

Section 7.2 continues step A_3 from phase 1. The Principle 3 (P 3)-System that I developed in Chap. 6 is compared with the "Rawlsian Core Precautionary Principle

T. Rechnitzer, *Applying Reflective Equilibrium*, Logic, Argumentation & Reasoning 27, https://doi.org/10.1007/978-3-031-04333-8_7

(RCPP)" (Gardiner 2006) and what I call the "Utilitarian Uncertainty Principle (UUP)" (Bognar 2011). The RCPP is an influential and widely cited PP proposal, so it makes sense to assess how it fares in the RE process. It also has been criticized: for example, the UUP is an explicit counter-proposal to the RCPP, with Bognar (2011) claiming that it can account for all the cases in the application set of the RCPP, but that it is also a defensible decision-principle in other cases—making the RCPP superfluous. As the comparison of the two candidates shows, this is not true—at least not with respect to my commitments. My own preliminary attempt at a candidate system, the P 3-System, is also rejected as a result of assessing it in comparison with the other two candidates, due to its lack of theoretical virtues.

After adjusting the commitments with respect to the RCPP in step B_3 (Sect. 7.3), I introduce a modified version of the RCPP in step A_4 (Sect. 7.4), the "Maximin-PP for combinations of uncertainty and incommensurability", drawing on the work of Aldred (2013). To have an additional sophisticated alternative, I adopt the proposal from Steel (2015) in the form of a "Tripartite Precautionary Approach (TPA)".

The Maximin-PP is selected because its higher ranking with respect to theoretical virtues arguably outweighs the slightly better account value of the TPA. However, even after adjusting the commitments with respect to the Maximin-PP in step B_4 (Sect. 7.5), substantial value-commitments remain unaccounted for (which would also be true if the TPA were chosen). I argue that this is because the Maximin-PP is a normative principle for *rational* choice, and that something more will be required to meet my pragmatic-epistemic objective of formulating a defensible *moral* precautionary principle. The latter is then the focus of Chap. 8.

Results from phase 2 are recapitulated and discussed in Sect. 7.6, which also includes Fig. 7.9, a schematic overview of the steps of phase 2. Throughout phase 2, gray boxes are again used to summarize the main points of each step. As before, only relevant or exemplary aspects of the process are described in detail, and readers can refer to Appendix A at the end of the book for the full list of commitments, candidates for (parts of) the system, background information, and case descriptions.

7.2 Step $A_{3.2}$: Comparing Principle 3-System, RCPP, and UUP

> Two candidates from the literature, the RCPP and the UUP, and my own candidate from phase 1, the "Principle 3-System" are comparatively assessed with respect to their ability to account for commitments (Sect. 7.2.1, see Fig. 7.1 for a summary) and their theoretical virtues (Sect. 7.2.2). In the overall comparison (Sect. 7.2.3), I argue that although not without problems, the RCPP is the most defensible candidate, and I adopt it as the current system at the end of step A_3.

Step A_3: Account for Commitments C_2

General Commitments on Precaution and Precautionary Decision-Making

	IC1	IC2	IC3	IC4	IC5	IC6	IC7	IC8	EC1	EC2	EC3	EC4	EC15	EC16	EC17	EC18	EC19	EC20
P3-System	?	× -2	-1.5	-1	✓ 6	✓ 6	-1	✓ 2	✓ 2	✓ 2	✓ 2	✓ 2	× -2	-1.5	✓ 4	~1	✓ 2	✓ 4
RCPP	?	× -2	-1.5	-1	✓ 6	✓ 6	-1	✓ 2	✓ 2	-0.5	-0.5	-0.5	× -2	-1.5	-1	✓ 2	× -2	× -4
UUP	?	× -2	-1.5	-1	× -6	-1.5	-1	× -2	✓ 2	-0.5	-0.5	-0.5	× -2	-1.5	-1	× -2	× -2	× -4

Commitments on Toy-Examples

	IC9	IC10	IC11	IC12	IC13	IC14	IC15	IC16	IC17	IC18	IC19	IC20	IC21	IC22
P3-System	✓ 6	✓ 6	✓ 4	?	✓ 6	✓ 6	✓ 4	✓ 4	✓ 4	✓ 6	✓ 4	✓ 6	✓ 4	✓ 6
RCPP	✓ 6	✓ 6	-1	?	✓ 6	✓ 6	-1	✓ 4	✓ 4	✓ 6	✓ 4	✓ 6	✓ 4	✓ 6
UUP	✓ 6	✓ 6	-1	?	✓ 6	✓ 6	✓ 4	✓ 4	-1	✓ 6	✓ 4	× -6	× -4	✓ 6

Commitments concerning Precaution, Climate Change, and Solar Radiation Management (SRM)

	IC23	IC24	IC25	IC26	IC27	IC28	IC29	IC30	IC31
P3-System	✓ 6	~ 2	✓ 4	✓ 4	✓ 4	✓ 6	✓ 4	-1	✓ 6
RCPP	✓ 6	~ 2	-1	✓ 4	✓ 4	-1.5	-1	-1	✓ 6
UUP	✓ 6	~ 2	✓ 4	✓ 4	✓ 4	✓ 6	✓ 4	✓ 4	✓ 6

Commitments on what counts as "Precautionary Measures"

	EC5	EC6	EC7	EC8	EC 9	EC10	EC11	EC12	EC13	EC14
P3-System	✓ 2	✓ 2	✓ 4	✓ 4	✓ 4	✓ 2	✓ 4	✓ 6	✓ 6	✓ 4
RCPP	-0.5	-0.5	-1	-1	-1	-0.5	-1	-1.5	✓ 6	-1
UUP	-0.5	-0.5	-1	-1	-1	-0.5	-1	-1.5	× -6	-1

+ NC 1 – NC 30

✓ full account	+2
~ partial account	+1
consistent, but no account	-0.5
× Conflict	-2
black: high weight	* 3
grey: medium weight	* 2
white: low weight	* 1

Account-Values: **P3-System: 164**
RCPP: 72
UUP: 35

Fig. 7.1 P3-System, UUP, and RCPP: account for commitments C_2

We are now resuming step A_3, which was started at the end of phase 1 in Chap. 6. In phase 1, I developed a candidate system, the Principle 3-System, which consists of Principle 3 and three additional parts.

Principle 3 (P 3) Where there are plausible threats of serious harm, precautionary measures that are proportional to the severity and plausibility of the threat should be taken.

P 3.1: Definition: Threat A threat is a possibility of harm that is uncertain.

P 3.2: Seriousness of Threats The seriousness of a threat is assessed according to (i) the potential for harm of the threat, and (ii) whether or not the possible harm is seen as reversible. [same content as IC 11]

P 3.3: *ExplicPrec* Explication of "Being a precautionary measure against an undesirable *x*": An action *a* is precautionary with respect to something undesirable *x* if *a* fulfills the following necessary and jointly sufficient criteria:

1. Intentionality: *a* is performed with the intention of preventing *x*.
2. Uncertainty: the agent does not believe it to be certain or highly probable that *x* will occur if *a* is not performed.
3. Reasonableness: the agent has externally good reasons (a) for believing that *x* might occur, (b) for believing that *a* will in fact at least contribute to the prevention of *x*, and (c) for not believing it to be certain or highly probable that *x* will occur if *a* is not performed.

I.e., Principle 3 is supposed to recommend those actions that fulfill the criteria of the explication, and are, on top of that, proportional to the severity and plausibility of the threat.

In the remainder of step A_3, this candidate system will be compared with the Rawlsian Core Precautionary Principle (RCPP) and the Utilitarian Uncertainty Principle (UUP). Gardiner's RCPP is an influential PP interpretation (Gardiner 2006), whereas the UUP is derived from Bognar's criticism and counter-proposal against the RCPP (2011). They have been selected based on the literature survey in Chap. 4.

The Rawlsian Core Precautionary Principle (RCPP) is a maximin-decision rule that is qualified through four jointly sufficient conditions. The first three conditions are cited (and slightly paraphrased) from Gardiner (2006, 74). The fourth condition is added on p. 51: "[The] RCPP needs some way of distinguishing a set of reasonable outcomes to contrast with those outcomes which are merely imaginable. This suggests that the three Rawlsian criteria mentioned so far must be supplemented with a further requirement: that the range of outcomes considered are in some appropriate sense "realistic," so that, for example, only credible threats are considered."

Rawlsian Core Precautionary Principle (RCPP) *If* four conditions are fulfilled:

1. **No Probabilities:** There is no, or no reliable, probability information about the possible outcomes available,
2. **Care Little for Potential Gains:** decision-makers care relatively little for potential gains that might be made above the minimum that can be guaranteed by the maximin approach,
3. **Unacceptable Outcomes:** the courses alternative to the one selected by maximin have unacceptable outcomes, and
4. **Reasonable Outcomes:** the range of outcomes considered are in some appropriate sense "realistic" or reasonable,

then decision-makers should choose the course of action with the best worst case.

The Utilitarian Uncertainty Principle consists of a combination of the principle of maximizing expected utility and the principle of indifference: it tells us to treat all outcomes as equally probable if no probability information is available, then to calculate their expected utilities, and to select the option/course of action that has the highest expected utility:

Utilitarian Uncertainty Principle (UUP) If no or no reliable probability information is available, treat all outcomes as equally probable, and choose the option that has the highest expected utility.

7.2.1 P 3-System, RCPP, and UUP: Account for Commitments

See Fig. 7.1 for a comparison of the Principle 3-System, the Rawlsian Core Precautionary Principle, and the Utilitarian Uncertainty Principle with respect to account for the current commitments C_2. Changes in commitments from C_1 to C_2, i.e., the emerging commitments EC 14–EC 20, are marked by different table borders.

The P 3-System reaches the highest account value out of the three, 164. The RCPP reaches 72, whereas the UUP only has an account value of 35 because it in fact conflicts with some strong commitments, for example:

IC 5 Don't risk great harm in pursuit of modest benefit. [high]

EC 13 Precautionary measures should be effective in preventing or substantially ameliorating either a threat or the harm of a threat. [high] [emerged at Step B_1]

The toy example *Disproportionate Outcomes 2* further helps to illustrate how the UUP conflicts with my commitments:

Case 8: Disproportionate Outcomes 2 We have to decide between two alternative courses of action, "Safe 2" and "Risky 2" (see Fig. 7.2).

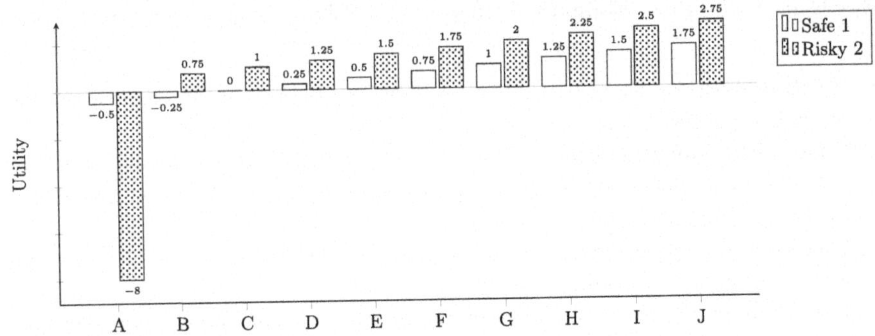

Fig. 7.2 Possible outcomes in Case 8, *Disproportionate Outcomes 2*

IC 20 In case 8, *Disproportionate Outcomes* 2, the option "Safe 2" should be chosen. [high]

IC 21 In case 8, *Disproportionate Outcomes* 2, the option "Safe 2" should be chosen **because** the worst case of option "Risky 2" is disproportionately worse than what we could gain from it as compared with "Safe 2". [medium]

The commitments concerning precautionary measures, EC 5–EC 14, are accounted for by the P 3-System via the explication of "being a precautionary measure against an undesirable *x*", and they are consistent with the RCPP and the UUP. However, the commitments on precautionary measures are only relevant for the P 3-System because it needs the explication of "being a precautionary measure against an undesirable *x*" in order to be able to yield somewhat determinate verdicts. The other two candidates yield determinate verdicts without this "detour" of first assessing whether a measure is precautionary or not. Still, there are two important differences between the two: while measures recommended by the RCPP will meet the criteria for being a precautionary measure, this is not the case for every measure the UUP will recommend, as, e.g., its failure to account for EC 13 shows.[1]

This is a point in favor of the RCPP, because as I argued in Sect. 6.8.2, not every precautionary measure is defensible. A justified PP should only recommend such measures that are warranted—but neither should it recommend measures that cannot even be classified as precautionary in the first place. Thus, the RCPP no longer needs the explication of precautionary measures, which was developed in the first phase of the case study (Chap. 6). We can thus move the explication to the background, as it is no longer needed as a part of the system.

[1] This commitment reads as follows:

EC 13 Precautionary measures should be effective in preventing or substantially ameliorating either a threat or the harm of a threat. [high] [emerged at Step B_1]

7.2.2 P 3-System, RCPP, and UUP: Theoretical Virtues

The three candidates are comparatively assessed and ranked with respect to their theoretical virtues. In the descriptions, I focus on the most salient features that distinguish the three candidates, setting aside those that they share. For example, all three candidates have to presuppose that we can distinguish plausible or reasonable outcomes that should be considered for the decision process from those that are not. This decreases their determinacy. But since all of them face the same problem, it does not make a difference for their ranking.

Determinacy The virtue of *determinacy* demands that "The target system should, together with relevant factual information, yield determinate verdicts, i.e., both its conditions of application and its verdicts should be precise and clear enough" (see Sect. 5.5 for more on this and the other virtues).

In specifying the kind of possible outcomes that "trigger" precautions, the RCPP is slightly more determinate than the P 3-System, since there is at least a "trivial", straightforward interpretation of "unacceptable" available, whereas "serious" leaves more room for discretion. But aside from this, "unacceptable" is still not very determinate without further specification, and it is also far from clear whether the trivial interpretation is the correct/best one.

The "care little for potential gains" criterion of the RCPP decreases its determinacy, since this criterion can be interpreted in several ways: does it mean that the minimum (best worst case) is already really good, so that additional gains don't mean much? Or that there is only a small difference between the best worst case and the best case?

On the other hand, the determinacy of P 3-System is impaired through reference to "irreversibility": the relevant sense of "irreversible" is far from clear, and it could be spelled out in a number of ways (Randall 2011, 57–60; 70–72).

Lastly, RCPP and UUP refer to probabilities without clarifying the relevant sense of *probability*, e.g., should we apply them when we have no objective probabilities available? This might often be the case. Or when we have no subjective probabilities available? That would be much more rare (Hansson 2008; Roser 2017, cf.). On the other hand, the P 3-System refers to threats which are possibilities of harm that are *uncertain*, which is similarly unclear.

As for the determinacy of verdicts, the RCPP and the UUP both yield determinate verdicts by singling out exactly one course of action that should be taken. The P 3-System, on the other hand, is not so clear-cut, since while it does include an explication of "precautionary measure", it does not further explicate when such a measure is *proportional* to the seriousness and plausibility of a threat. The use of this concept relies thus on our existing (pre-theoretical) understanding of proportionality, which leaves substantial room for discretion.

Based on this assessment, I rank the three candidates as following with respect to their determinacy:

$$UUP > RCPP > P3$$

Practicability A candidate system has the virtue of *practicability* when it is applicable in the sense that it specifies relevant information about actions and other items of evaluation that human beings can typically obtain and use to arrive at moral verdicts. E.g., it should process inputs that are typically available to us, and yield verdicts that are realizable by us. For more explanation, see p. 116.

The problem with the practicability of the P 3-System is that its determinacy is so low that it is not really possible to assess how accessible the relevant information typically is for us—because it is unclear what the relevant information would be.

For the RCPP, we have to be able to rank outcome values ordinally, and to be able to identify the worst case of an option. While the first speaks for its practicability—ordinal rankings are much easier to obtain than rankings on interval or ratio scales—the latter is not very practicable as long as we are missing a clear criterion for "reasonable" outcomes.

The UUP seems more practicable than both RCPP and P 3-System. Its main problem as regards practicability is that we need be able to assign cardinal utilities to the outcomes in order to calculate expected utilities, i.e., we need to be able to rank outcome utilities on interval scales. This might not always be possible.

I rank the practicability of the three candidates as following:

$$UUP > RCPP > P3$$

Broad Scope A candidate system has a broad scope when it has a broad *range of applicability*, i.e., it should tell us in as many cases as possible whether or not (which specific) precautionary measures are required. For more on scope, see p. 116.

The range of applicability of the RCPP consists of all situations in which outcomes are in some relevant sense reasonable, decision-makers care little about potential gains that can be made above the minimum that can be guaranteed by following maximin, no (reliable) probability information is available, and alternatives to the option recommended by maximin all include unacceptable outcomes. This is the smallest range of applicability of the three candidates.

Then we have the UUP, which covers all situations where values of reasonable outcomes can be ranked on cardinal scales.

The P 3-System can also handle situations when outcomes are incommensurable, as opposed to the UUP, where such situations are not in its range of applicability. I thus rank the three candidates according to their scope as following:

$$P3 > UUP > RCPP$$

Simplicity I understand the theoretical virtue of *simplicity* as demanding that the conceptual apparatus of the target system should be economical in the sense that the concepts it includes that cannot be reduced to each other are kept to a minimum. See also Sect. 5.5.

The P 3-System includes thirteen different concepts: threat; plausible (threat/outcome); serious (threat/harm); proportionality (of precautionary measures to plausibility and severity of a threat); potential for harm (of a threat);

(ir)reversibility (of harm); prevention; intention; belief; certain; highly probable; externally good reasons (for believing something); contributing to the prevention of an event.

The theoretical apparatus of the RCPP consists of seven concepts: probability; outcome; course of action/option; maximin; "to care little for potential gains above a specific guaranteed minimum"; to be unacceptable; reasonable outcomes. The technical apparatus of the UUP consists of five concepts: probability; outcomes; option/course of action; utility; expected utility.

By including thirteen different concepts, the P 3-System does not seem to be particularly simple, even if considered on its own. The RCPP is considerably simpler in only including seven concepts, while the UUP is even simpler with only five. Consequently, the ranking of the three candidates according to their simplicity is:

$$UUP > RCPP > P3$$

Overall Ranking: Theoretical Virtues of P 3, RCPP, and UUP Here is a short overview of how the three candidates fare overall with respect to the theoretical virtues.

Determinacy:	$UUP > RCPP > P3$
Practicability:	$UUP > RCPP > P3$
Scope:	$P3 > UUP > RCPP$
Simplicity:	$UUP > RCPP > P3$

The UUP is always better than the RCPP, so we get overall $UUP > RCPP$. The RCPP is always better than the P 3-System *aside from* scope, where the P 3-System even outranks the UUP. How to handle this trade-off is discussed in the next subsection, where the candidates are compared overall, i.e., with respect to both account and theoretical virtues.

7.2.3 Overall Comparison of P 3, RCPP, and UUP

The P 3-System has the broadest scope of the three candidates, and also the highest account value (164). On the other hand, it ranks last with respect to the virtues of determinacy, practicability, and simplicity. We also have to consider that its high rank with respect to account (a) rests on the assumption that all outcomes are plausible in the relevant sense, and (b) is also partly due to the fact that it can simply account for some of the commitments because they are also part of the P 3-System (e.g., EC 2–4). I argue that compared with the available alternatives, this trade-off between account and theoretical virtues is not defensible: we should reject the P 3-System as a candidate at this point.[2]

With the UUP and the RCPP, we are also facing a trade-off between account and theoretical virtues: while the UUP ranks higher with respect to overall theoretical virtues than the RCPP, the latter has a higher account value (RCPP: 72, UUP: 35, see Fig. 7.1, p. 155). The low account value of the UUP is partly due to its conflict with some central commitments that have a high weight, e.g. IC 5, EC 13, and IC 20, and others with a medium weight, e.g., IC 21. And as the assessment of theoretical virtues has shown, even though the UUP ranks highest with respect to *determinacy* and *scope*, it is not completely unproblematic.

Thus, I argue that the theoretical virtues of the UUP cannot compensate for its low account value, and that for now, the RCPP is the most convincing candidate, despite its problems. I thus adopt the RCPP as the current system.

7.3 Step B3: Adjusting Commitments to the RCPP

> Commitments are adjusted with respect to the current system, the RCPP. In Sect. 7.3.1, I discuss whether the set of current commitments can be adjusted to increase agreement with the current system. The adjustments of excluding from the subject matter (i) cases with probabilities, or (ii) with more than one acceptable worst case/without any acceptable worst case, are rejected as not defensible. After searching for further relevant commitments in Sect. 7.3.2, the current set of commitments at the end of step B3 is described in Sect. 7.3.3 and summarized in Fig. 7.6.

At the end of the last step, the Rawlsian Core Precautionary Principle (RCPP) was chosen as the current system. However, the assessment and comparison of the three candidate systems in step A3 has shown that even though the RCPP is currently the most defensible candidate, it still has significant room for improvement. When adjusting the current commitments with respect to the RCPP, we have to keep this in mind and should not give too much weight to the current system.

[2] We could also try to further systematize the aspects of the P 3-System, etc.—this would not pose a principled problem for the RE application, and it could be done similarly to how the P 3-System was constructed in phase 1. However, this also makes it not very interesting from the perspective of the RE case study, especially since more developed candidates are available to be used in the RE steps.

7.3.1 *Trying to Increase Account*

First, I discuss whether commitments that are in conflict with the RCPP should be adjusted (a). Then, I discuss whether cases where either (b) probabilities are available or (c) cases that have more than one acceptable worst case, respectively no acceptable worst case, should be excluded from the subject matter to increase account, since the RCPP does not yield verdicts in such cases. Lastly, some commitments are clarified (d).

(a) Adjusting Conflicting Commitments The RCPP conflicts with four commitments that are all general commitments concerning precaution and precautionary decision-making, not case-specific judgments.

First, we have the commitments to the Wingspread PP:

IC 2 When an activity raises threats of harm to human health or the environment, precautionary measures should be taken even if some cause-and-effect relationships are not fully established scientifically. (Wingspread Formulation of the Precautionary Principle) [low]

There can be cases where IC 2 demands that measures should be taken whereas RCPP recommends against them: for example, if there is a course of action that has a small harm to the environment in its set of possible outcomes, but another alternative course of action has an unacceptably huge economical loss in its set of outcomes, then RCPP will recommend the first alternative, given that the other criteria of the RCPP are fulfilled. That is, RCPP will recommend the action that entails a threat to the environment when otherwise we would face unacceptable economic loss, but IC 2 only refers to threats of harm to human health or the environment, and does not take, e.g., economic loss into account.

I argue that interpreting the Wingspread PP in such an absolutist sense is not very plausible, and propose to replace IC 2 by the following new commitment:

C 2 When an activity raises threats of harm to human health or the environment, then, *pro tanto*, precautionary measures should be taken even if some cause-and-effect relationships are not fully established scientifically. [medium] [replaced IC 2 at Step B3]

Making the claim *pro tanto* means that it is still respected that threats to the environment and/or human health warrant precaution, but it also takes into account that this claim sometimes can be overridden. I thus argue that it is still close enough to the original input commitment to be seen as respecting it. While the RCPP does not account for C 2, it is at least consistent with it, which leads to a slight increase of the agreement between commitments and current system.

The next current commitment the RCPP conflicts with is:

EC 15 Precautionary measures should not introduce serious threats of their own. [low] [emerged at Step B2]

The RCPP only refers to unacceptable threats, and if we assume that while all threats of unacceptable outcomes are serious, but not all serious threats are threats of unacceptable outcomes, then it might also happen that the precautionary measure that it endorses will introduce a serious threat. However, independently of the RCPP, demanding that a precautionary measure is never allowed to introduce serious threats is potentially paralyzing, as it is possible that there is no such option. I thus propose the following replacement for EC 15, and argue that the RE criterion of "respecting input commitments" is fulfilled:

C 3 Precautionary measures should not introduce threats that are equally or more severe than the threats they are aimed at, i.e., threats that have the same or a greater potential for harm. [medium] [replaced EC 15 at Step B_3]

I argue that the remaining two conflicting commitments cannot defensibly be adjusted:

EC 19 The price of precaution should be proportional to the seriousness and the plausibility of the threat, given the available alternatives. [low] [emerged at Step B_2]

The RCPP conflicts with this because once no probabilities are available, all plausible outcomes are treated the same, i.e., additional comparisons of plausibility do not play a role. However, I think that it makes sense to somehow take into account our epistemic state concerning the possible outcomes, and that there can be comparisons made even if we have no probability information available. I thus argue that rejecting this commitment would not fulfill the RE criterion of "respecting input commitments".

EC 20 The target PP applies to plausible and serious threats and prescribes measures that are proportional to the severity and plausibility of the threat. [medium] [emerged at Step B_2]

The RCPP does not apply to all serious threats, given that not all of them will also be threats of unacceptable outcomes. The RCPP also fails to account for the part of EC 20 that concerns plausibility: all we have to know about outcomes in order to apply the RCPP is that they are "reasonable", but we do not know their probability. Further information is not considered, neither about threats nor about precautionary measures against them—the RCPP does not include any comparisons of plausibility of outcomes. Consequently, the proportionality of measures to the plausibility of a threat does not play a role: the RCPP conflicts with the constraints EC 20 puts on the target PP. Should we reject EC 20 on this basis? I argue that the answer should be no, based on a similar argument as for EC 19: comparisons of plausibility are possible and sensible even if we have no probability information available.

The problem for the RCPP that is bigger than conflicting commitments is the broad range of commitments on which it remains silent because it does not apply. It is unable to account for any commitments on cases in which probabilities are available, or in which more than one of the available courses of action has an acceptable worst case, or in which no available course of action does have an acceptable worst case. Instead of zooming in on each of the particular, case-specific commitments, for each of these groups I discuss whether or not it makes sense to exclude them and thereby increase the RCPP's ability to account for the current commitments.

(b) Excluding Probabilities from Subject Matter? The RCPP does not apply to cases where probabilities are available. Consequently, commitments concerning such cases are not in its application set. However, it is not uncommon to restrict the scope of PPs to situations in which we know the possible outcomes, but not their probabilities—i.e., to restrict it to situations of *decision-theoretic uncertainty*. The idea behind this seems to be that as long as we do not only know the range of possible outcomes, but also their probabilities—i.e., as long as we are in a situation of *decision-theoretic risk*—we do not need something like a PP because we already know what to do: we can calculate expected utilities and choose the course of action that maximizes expected utility.

However, I think that there is a strong argument that there are at least some cases in which we should take precautionary measures even in situations of decision-theoretic risk. Choosing the course of action that maximizes expected utility might only make sense when potential losses are insurable, when we know that we can try several times, and/or when potentially bad outcomes are not extremely severe.

Randall (2011) identified relevant cases in which precaution is relevant—i.e., that a PP should be able to account for—even though probabilities are available.

For example, let us compare two courses of action. Option *a*, "widely dispersed", has extremely good but also extremely bad possible outcomes. The outcome distribution is normal, i.e., symmetric around the mean, median, and mode; and the expected value is positive. Option *b*, "compact", has the same expected utility as *a*, but its outcomes are much more narrowly dispersed, i.e., its best and worst cases are much less extreme, while the mode—the most likely outcome—is much more likely than in *a*. The two alternatives are illustrated in Fig. 7.3.

And things get worse when outcomes are not distributed along a normal distribution, but in some disproportional and/or asymmetric way. The set of possible

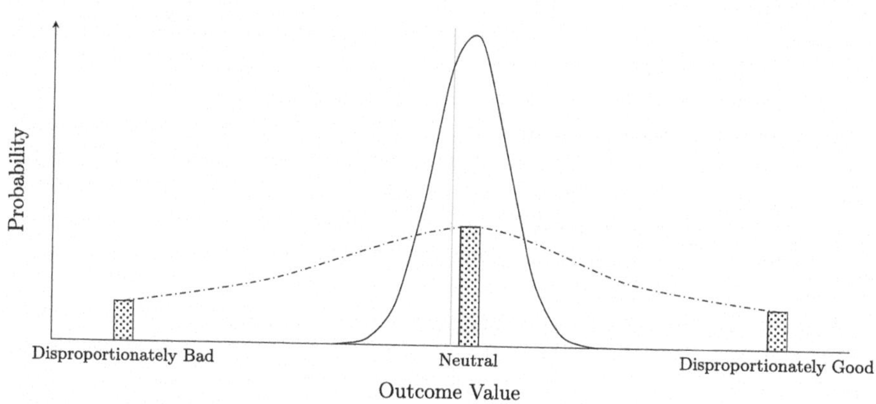

Fig. 7.3 Randall (2011, 112): Catastrophic possibilities with a normal outcome distribution and highly dispersed outcome possibilities vs. a compact normal distribution

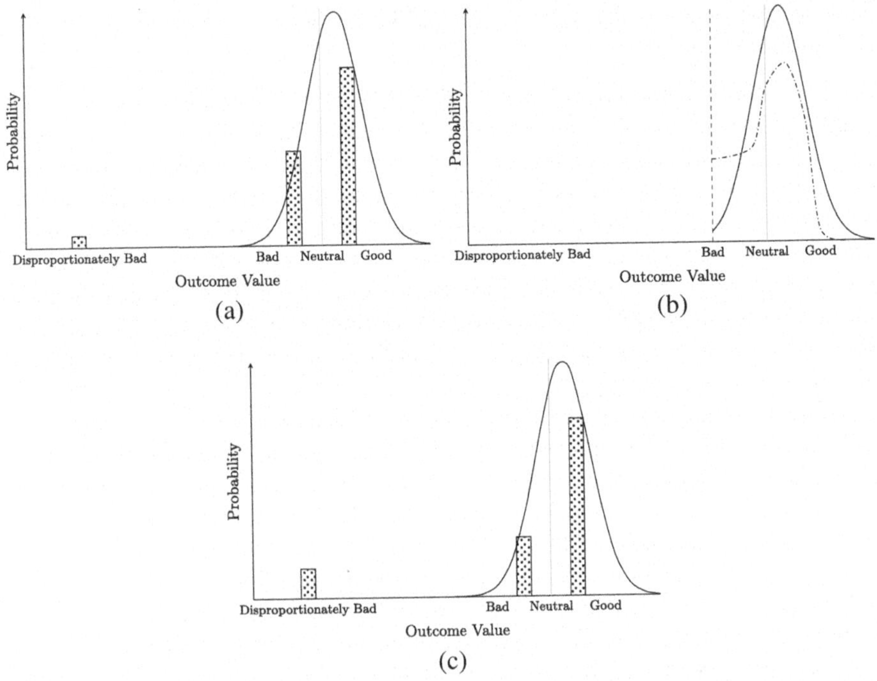

Fig. 7.4 (**a**) Disproportionate threat of harm. (**b**) Asymmetric threat of harm. (**c**) Disproportionate and Asymmetric Threat of Harm (graphics based on Randall 2011, 113, printed with permission from Cambridge University Press)

outcomes of a course of action involves a **disproportionate threat of harm** when it includes possibilities of harm that, compared with the expected value of the course of action or with its most likely outcome, are disproportionately bad. An example is shown in Fig. 7.4a, where three discrete outcomes are possible, with a modest net gain being the most likely outcome, closely followed in likelihood by a modest net harm. Even though this extreme harm is unlikely, its possibility is non-trivial. And in comparison with what we can gain, it seems clearly disproportionate.

Asymmetric threats of harm is illustrated in Fig. 7.4b by a continuous distribution of outcome possibilities that is truncated at outcomes that are clearly bad but not disproportionately so (Randall 2011, 114). "Asymmetric" refers to the distribution of likelihoods, meaning that substantial (even though not necessarily disproportional) harm is more likely than it would be were the distribution of outcome likelihoods symmetric.

A threat is **disproportionate and asymmetric** when a disproportionately bad outcome is asymmetrically likely (Randall 2011, 114). This is illustrated in Fig. 7.4c.

All in all, while there might be cases where we know the probabilities of the possible outcomes and no special precaution is required—like in the case of the

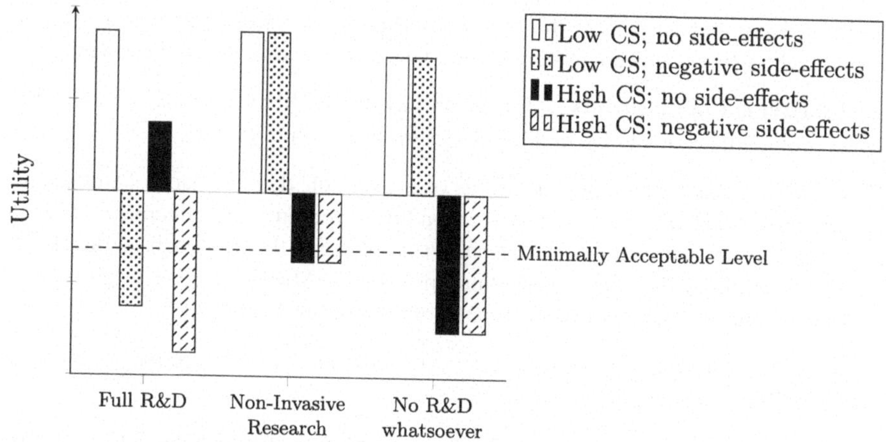

Fig. 7.5 Possible Outcomes in Case 3, *R&D into SRM, Two Kinds of Research*

compact normal distribution—this is not true for all situations in which we have probabilities. Excluding them from the subject matter can thus not be defended.

(c) Excluding Cases with More Than One/without Any Acceptable Worst Case? Because of condition 2 of the RCPP, "decision-makers care relatively little for potential gains that might be made above the minimum that can be guaranteed by the maximin approach", the worst case selected by the maximin-rule—i.e., the best worst case—has to be at least acceptable. But according to condition 3, "the courses alternative to the one selected by maximin have unacceptable outcomes". This means that the RCPP cannot account for any commitments where more than one course of action has an acceptable worst case, nor for commitments about cases where none of the courses of action has an acceptable worst case.

The latter is for example the case in case 3, *R&D into SRM, Two Kinds of Research*, where none of the three available courses of action has a worst case that is above the minimally acceptable level. The specific details of Case 3 can be found in the appendix, here only the distribution of outcomes is important, which can be seen in Fig. 7.5. "CS" refers to climate sensitivity, and the negative side effects refer to whether or not solar radiation management would work as intended as a climate engineering technology.

We could, e.g., try to argue that if there is no "safe" option, in the sense that even if everything goes wrong, we still end up with an acceptable outcome, then precaution is no longer an option and we might have to look to other decision criteria. But at this moment, it seems a bit premature to exclude these cases without even trying to explore whether there is another candidate system that can account for them.

As for whether to exclude cases where more than one worst case is acceptable, this seems even less plausible to me. At the very least, the target PP should leave it open to us in such a situation which course of action we want to choose.

So I argue that neither of the identified classes of commitments that are not accounted for by the current system should be excluded from the subject matter.

(d) Clarifying Commitments It is unclear how the RCPP relates to IC 1 and IC 24, but arguably this is due to the lack of clarity of those commitments:

IC 1 Where there are threats of serious or irreversible damage, lack of full scientific certainty shall not be used as a reason for postponing cost-effective measures to prevent environmental degradation. (Principle 15 of the Rio Declaration) [low]
IC 12 The claim that the measles/mumps/rubella (MMR) vaccine might cause autism is not a reason not to vaccinate your child. [medium]

I propose to adjust them in the following way, which helps to determine whether or not the RCPP can account for them:

C 1 When there is a plausible threat of serious or irreversible harm to the environment, then uncertainty of the harm must not lead to postponing cost-effective measures to prevent it. [medium] [replaced IC 1 at Step B3]
C 4 The claim that the measles/mumps/rubella (MMR) vaccine might cause autism does not constitute a plausible threat. [medium] [replaced IC 12 at Step B3]

These adjustments are defensible as respecting the input commitments, since they are intended as clearer formulations replacing the original, more vague, ones.

However, they are not accounted for by the RCPP, which is the current system, but are both merely consistent with it: since the RCPP does not say anything about how to identify plausible threats, it does not tell us anything about C 4. And while the RCPP does not tell us that we should postpone measures in the face of uncertain harm, i.e., it does not conflict with C 1, neither does it tell us that measures must not be postponed. The RCPP sometimes can be used to comply with the demand of C 1, i.e., to identify measures to prevent serious or irreversible harm to the environment, but it does not itself account for this demand.

7.3.2 *Searching for Further Relevant Commitments*

When discussing whether cases with probabilities should be excluded from the subject matter (Sect. 7.3.1), I argued that there are cases where probabilities are available, but taking precautions is warranted. This revealed further relevant commitments, and these should be added as emerging input commitments:

EC 21 The wider the dispersal of the outcome value distribution of a course of action, the more precaution is warranted by its negative outcomes. [medium] [emerged at Step B3]
EC 22 *Pro tanto*, cases that involve threats of disproportionate harm warrant precautionary measures, even if this harm is very unlikely. [medium] [emerged at Step B3]

7.3.3 *The Adjusted Set of Current Commitments, C_3*

In Fig. 7.6, you can see the adjusted set of commitments compared with the account table from step A_3, Fig. 7.1. For a full list of the text of each commitment, you can, as always, refer to Appendix A at the end of the book. The commitments that have been adjusted with respect to the current system are marked with double border cells. These adjustments slightly increase the account value between RCPP and the current commitments from 72 to 78.

7.4 Step A_4: From RCPP to Maximin-PP

By adjusting some of the weaknesses of the RCPP, the "Maximin-PP for Combinations of Uncertainty and Incommensurability (Maximin-PP)" is introduced as an additional candidate system (based on Aldred 2013). In order to increase the competition, a slightly adapted version of the PP proposal by Steel (2015) is introduced as the "Tripartite Precautionary Approach (TPA)". After comparing five candidate systems with respect to account (Sect. 7.4.1) and their theoretical virtues (Sect. 7.4.2), the Maximin-PP is chosen in the overall comparison (Sect. 7.4.3) against the TPA, which is a close runner-up.

To adjust the current system, I again compare different candidate systems. Even though the Principle 3-System (P 3) and the Utilitarian Uncertainty Principle (UUP) were rejected at step A_3, they are again considered as potential alternatives to the RCPP, this time with respect to the current commitments C_2. However, learning from the weaknesses of the current system, the RCPP, I also assess two further candidate systems. Firstly, I propose an adjustment of the RCPP that has a wider scope and should also be more determinate, the "Maximin-PP for Combinations of Uncertainty and Incommensurability (Maximin-PP)". Secondly, I introduce a slightly adapted version of Steel's (2015) proposal for a PP, which I call the "Tripartite Precautionary Approach (TPA)".

Introducing the Maximin-PP as an Adjustment of the RCPP The small scope, low determinacy, and low practicability of the Rawlsian Core Precautionary Principle (RCPP) remains a problem. A candidate that would score higher with respect to these virtues than the RCPP (that preferably includes the RCPP in its scope) is desirable. In order to find such a candidate, let us have a closer look at the weaknesses of the RCPP.

For the RCPP, the main problem with respect to determinacy is its condition 2, which demands that "decision makers care relatively little for potential gains that might be made above the minimum that can be guaranteed by the maximin

Step B_3: Current Set of Commitments C_3

General Commitments on Precaution and Precautionary Decision-Making

	C1	C2	IC3	IC4	IC5	IC6	IC7	IC8	EC1	EC2	EC3	EC4	C3	EC16	EC17	EC18	EC19	EC20	EC21	EC22
P3-System			-1.5	-1	✓6	✓6	-1	✓2	✓2	✓2	✓2	✓2		-1.5	✓4	~1	✓2	✓4		
RCPP	-0.5	-0.5	-1.5	-1	✓6	✓6	-1	✓2	✓2	-0.5	-0.5	-0.5	✓4	-1.5	-1	✓2	x -2	x -4	-1	-1
UUP			-1.5	-1	x -6	-1.5	-1	x -2	✓2	-0.5	-0.5	-0.5		-1.5	-1	x -2	x -2	x -4		

Commitments on Toy-Examples

	IC9	IC10	IC11	C4	IC13	IC14	IC15	IC16	IC17	IC18	IC19	IC20	IC21	IC22
P3-System	✓6	✓6	✓4		✓6	✓6	✓4	✓4	✓4	✓6	✓4	✓6	✓4	✓6
RCPP	✓6	✓6	-1	-1	✓6	✓6	-1	✓4	✓4	✓6	✓4	✓6	✓4	✓6
UUP	✓6	✓6	-1		✓6	✓6	✓4	✓4	-1	✓6	✓4	x -6	x -4	✓6

Commitments concerning Precaution, Climate Change, and Solar Radiation Management (SRM)

	IC23	IC24	IC25	IC26	IC27	IC28	IC29	IC30	IC31
P3-System	✓6	~2	✓4	✓4	✓4	✓6	✓4	-1	✓6
RCPP	✓6	~2	-1	✓4	✓4	-1.5	-1	-1	✓6
UUP	✓6	~2	✓4	✓4	✓4	✓6	✓4	✓4	✓6

✓ full account	+2
~ partial account	+1
consistent, but no account	-0.5
x Conflict	-2
black: high weight	* 3
grey: medium weight	* 2
white: low weight	* 1

Fig. 7.6 End of Step B_3: Current Commitments C_3

approach". It is unclear what this means, exactly: that outcomes should be distributed in a way that the best worst case is almost as good as the overall best case? As an empirical claim about preferences of decision-makers? That in all alternative courses of action, potential gains that are above the overall best worst case are outweighed by their negative outcomes?

Furthermore, even if the criterion that rejected alternatives that have "unacceptable" outcomes is supposed to be a thick ethical claim, i.e., that to determine whether or not it is met we need to make value judgments and cannot describe it in merely factual terms, it still seems that a bit more could be said on what exactly makes an outcome unacceptable.

The scope of the RCPP is a problem, too, especially because its small range of applicability prevents it from accounting for whole classes of commitments. At this point, it seems reasonable that the target PP should be restricted to cases in which only reasonable/plausible outcomes are considered, and in which outcomes are at least comparable (even if not necessarily commensurable). But it would be desirable if we could broaden the scope of the target system beyond the RCPP's other restrictions to situations in which no probabilities are available, all courses of action alternative to the one selected by maximin include unacceptable outcomes, and "decision-makers care relatively little about potential gains that could be made above the best worst case".

What we want is ideally a new candidate system that scores better than the RCPP with respect to the virtues of practicability, scope, determinacy, as well as with respect to the account value; while being equal or at least similar to it with respect to simplicity (or maybe even simpler than the RCPP, too).

Starting with the conditions of application of the RCPP, is there a way to broaden their scope while at the same time rendering them more precise?

Aldred (2013) has made a proposal that looks very promising: by drawing on decision theory analyzing decisions under uncertainty, i.e., where it is not possible to attach one discrete probability to each outcome, he not only provides support for the specific criteria of the RCPP, but also argues that those criteria are just one configuration of features a decision problem can have that will justify following the maximin rule. Put differently, he proposes a PP candidate that is a qualified maximin rule, like the RCPP, but the criteria of this maximin-PP are more "flexible", i.e., they cover a broader scope. More specifically, he proposes a maximin-PP that is limited to specific combinations of uncertainty and incommensurability.

Since there are different interpretations of what it means for outcomes to be incommensurable, let me cite Aldred's (2013, 133) understanding: "Outcomes are incommensurable when, even in conditions of certainty, their value cannot be precisely measured along some common cardinal scale. In contrast, outcomes are incomparable when they cannot even be ranked on an ordinal scale. Thus comparability is a necessary but not sufficient condition for commensurability."

Maximin-PP for Combinations of Uncertainty and Incommensurability (Maximin-PP) Select the course of action with the best worst case if you are either:

- In a situation of decision-theoretic risk or uncertainty (or some combination), and the outcomes of the available actions can be ranked on an ordinal scale, and all courses of action alternative to the one selected by maximin have outcomes that are incommensurably worse than the best worst case; or
- In a situation of (partial) decision-theoretic uncertainty, outcomes can be ranked on a cardinal scale, and all courses of action alternative to the one selected by maximin have negative outcomes that outweigh every potential gain that could be made above the level that can be guaranteed by maximin.

The Tripartite Precautionary Approach Since P 3-System, RCPP, and UUP have already been assessed in detail, I am adding an additional candidate from the literature, which can be assessed and compared with them. I adapt the proposal from Steel (2015), which is an elaborate and comprehensive candidate. I call it the "Tripartite Precautionary Approach", since it provides a framework consisting of three main elements, which allows us to formulate specific *versions* of a precautionary principle.[3]

The **Tripartite Precautionary Approach (TPA)** consists of (cf. Steel 2015, 9–10):

- The **Meta Precautionary Principle (MPP)**: Uncertainty must not be a reason for inaction in the face of serious threats.
- The **Precautionary Tripod**: The elements that have to be specified in order to obtain an action-guiding precautionary principle version: If there is a threat that meets the *harm condition* under a given *knowledge condition* then a *recommended precaution* should be taken.
- **Proportionality**: Demands that the elements of the Precautionary Tripod are adjusted proportionally to each other, understood as *Consistency*: The recommended precaution must not be recommended against by the same PP version, and *Efficiency*: Among those precautionary measures that can be consistently recommended by a PP version, the least costly one should be chosen.

  The **strategy to obtain a PP decision rule by adjusting the precautionary tripod**: (1) select a desired safety target and define the harm condition as a failure to meet this target, (2) select the least stringent knowledge condition that results in a consistently applicable version of PP given the harm condition. To comply with the MPP, uncertainty must neither render the PP version inapplicable nor lead to continual delay in taking measures to prevent harm (cf. Steel 2015, 10).

[3] Steel calls it a "precautionary principle", but during a Workshop in Bern in May 2017 he said that "approach" might be more appropriate, and that he mostly chose "principle" to stick with the established terminology.

7.4.1 *Maximin-PP, TPA, P3, RCPP, and UUP: Account for Current Commitments*

In Fig. 7.7, the results from assessing account for the five candidate systems with respect to the current commitments C_3 is summarized. As before, the full list of commitments can be found in Appendix A. The TPA reaches the highest account value, 150, followed by the P 3-System with 141.5, the Maximin-PP with 127.5, the RCPP with 78, and lastly the UUP with 42.

7.4.2 *Theoretical Virtues of Maximin-PP, TPA, P3, RCPP, and UUP*

Since the theoretical virtues of the P 3-System, the RCPP, and the UUP have already been assessed in step A_1, I focus now on the Maximin-PP and the TPA. The rankings of the candidates with respect to theoretical virtues are described starting on p. 176, right before the overall comparison with respect to both theoretical virtues and account (7.4.2).

Determinacy The virtue of *determinacy* demands that: "The target system should, together with relevant factual information, yield determinate verdicts, i.e., both its conditions of application and its verdicts should be precise and clear enough" (see Sect. 5.5 for more on this virtue).

The P 3-System, the RCPP, and the UUP are ranked according to their determinacy as $UUP > RCPP > P3$. How do the Maximin-PP and the TPA compare?

I rank the Maximin-PP as on a par with the UUP:[4] both draw on existing decision theory, which means that the concepts they use are mostly already defined. The TPA, on the other hand, departs more from decision theory, which means that while most of its concepts are relatively clearly defined, there can be boundary cases for, e.g., what does count as uncertain, how to assess stringency of knowledge conditions, etc.—but I would still rank it higher than the RCPP with its notoriously unclear "care little for potential gains" criterion. We thus get the following ranking with respect to the virtue of determinacy:

$$UUP = MaximinPP > TPA > RCPP > P3$$

Practicability A candidate system has the virtue of *Practicability* when it is applicable in the sense that it specifies relevant information about actions and other items of evaluation that human beings can typically obtain and use to arrive at moral

[4] I am using "on a par" here instead of "equally good" to indicate that the UUP might be more determinate in some aspects and the Maximin-PP in others, i.e., they might not be equally determinate in all aspects.

Step A₄: Account for Commitments C₃

General Commitments on Precaution and Precautionary Decision-Making

	C1	C2	IC3	IC4	IC5	IC6	IC7	IC8	EC1	EC2	EC3	EC4	C3	EC16	EC17	EC18	EC19	EC20	EC21	EC22
P3-System	✓2	-0.5	-1.5	-1	✓6	✓6	-1	✓2	✓2	✓2	✓2	✓2	✓4	-1.5	✓4	~1	✓2	✓4	✓4	✓4
RCPP	-0.5	-0.5	-1.5	-1	✓6	✓6	-1	✓2	✓2	-0.5	-0.5	-0.5	✓4	-1.5	-1	✓2	× -2	× -4	-1	-1
UUP	-0.5	-0.5	-1.5	-1	× -6	-1.5	-1	× -2	✓2	-0.5	-0.5	-0.5	× -4	-1.5	-1	× -2	× -2	× -4	× -4	-1
Maximin-PP	-0.5	-0.5	-1.5	-1	✓6	✓6	-1	✓2	✓2	-0.5	-0.5	-0.5	✓4	-1.5	-1	✓2	✓2	✓4	✓4	✓4
TPA	✓2	✓2	✓6	-1	✓6	✓6	-1	✓2	✓2	✓2	-0.5	✓2	✓4	-1.5	-1	✓2	✓2	✓4	✓4	✓4

Commitments on Toy-Examples

	IC9	IC10	IC11	C4	IC13	IC14	IC15	IC16	IC17	IC18	IC19	IC20	IC21	IC22
P3-System	✓6	✓6	✓4	-1	✓6	✓6	✓4	✓4	✓4	✓6	✓4	✓6	✓4	✓6
RCPP	✓6	✓6	-1	-1	✓6	✓6	-1	✓4	✓4	✓6	✓4	✓6	✓4	✓6
UUP	✓6	✓6	-1	-1	✓6	✓6	✓4	✓4	-1	✓6	✓4	× -6	× -4	✓6
Maximin-PP	✓6	✓6	✓4	-1	✓6	✓6	✓4	✓4	✓4	✓6	✓4	✓6	✓4	✓6
TPA	✓6	✓6	✓4	-1	✓6	✓6	✓4	✓4	✓4	✓6	✓4	✓6	✓4	✓6

Commitments concerning Precaution, Climate Change, and Solar Radiation Management (SRM)

	IC23	IC24	IC25	IC26	IC27	IC28	IC29	IC30	IC31
P3-System	✓6	~2	✓4	✓4	✓4	✓6	✓4	-1	✓6
RCPP	✓6	~2	-1	✓4	✓4	-1.5	-1	-1	✓6
UUP	✓6	~2	✓4	✓4	✓4	✓6	✓4	✓4	✓6
Maximin-PP	✓6	~2	✓4	✓4	✓4	✓6	✓4	-1	✓6
TPA	✓6	~2	✓4	✓4	✓4	✓6	✓4	✓4	✓6

✓ full account	+2
~ partial account	+1
consistent, but no account	-0.5
× Conflict	-2
black: high weight	* 3
grey: medium weight	* 2
white: low weight	* 1

Account-Values:
P3-System: 141.5
RCPP: 78
UUP: 42
Maximin-PP: 127.5
TPA: 150

Fig. 7.7 Comparing candidate systems with respect to the current commitments C_3

verdicts. That is, it should process inputs that are typically available to us, and yield verdicts that are realizable by us. For more explanation, see p. 116.

While incommensurability as part of the Maximin-PP is relatively clearly defined, in specific cases there might still be insecurity about what does count as incommensurable and how we determine this, especially when more than one agent is concerned.

We also have the problem of how to identify the worst cases of courses of action: because for every course of action we can fabricate some possible catastrophic outcome, we need some criterion for where we should draw the line concerning what does and what does not count as a reasonable outcome.

But aside from that, it seems that the Maximin-PP processes inputs that are relatively easily accessible for us, e.g., ordinal rankings of outcomes, seem relatively unproblematic; and once the relevant sense of probability is settled, it should also be accessible in which epistemic situation we find ourselves (i.e., decision-theoretic risk versus uncertainty).

For the practicability of the TPA, the required ranking of harm conditions, respectively precautionary measures, with respect to knowledge conditions might not always be very practicable. Also, identifying precautionary measures can be difficult, if a PP version has to be formulated and adjusted to each new combination of harm and knowledge conditions. We might be able to settle whether or not a threat is serious in most cases, even though the TPA itself does not provide any guidelines for it.

All in all, the TPA is relatively practicable in the sense I use here. However, I argue that it presupposes more information than the Maximin-PP and requires more steps before an action-guiding decision results, and I therefore rank the TPA as less practicable than the Maximin-PP.

The P 3-System, the RCPP, and the UUP are ranked with respect to the virtue of practicability as $UUP > RCPP > P\,3$. I have already established that the TPA is less practicable than the Maximin-PP. I argue that both Maximin-PP and TPA are more practicable than the RCPP, but how do they rank with respect to the UUP? I consider the TPA to be less practicable than the UUP because it requires more information and more adjustment. The Maximin-PP is more practicable because it requires only ordinal ranking of outcomes where the UUP requires cardinal rankings, and otherwise it is not less practicable.

Thus, with respect to practicability, we get the following ranking:

$$MaximinPP > UUP > TPA > RCPP > P3$$

Broad Scope A candidate system has a broad scope when it has a broad *range of applicability*. I.e., it should tell us in as many cases as possible whether or not (which specific) precautionary measures are required. For more on scope, see p. 116.

When ranking the scope of the P 3-System, the UUP, and the RCPP, we got the following: $P3 > UUP > RCPP$

I argue that the TPA has the same range of applicability as the P 3-System (all situations with and without serious threats). And the Maximin-PP, too, is applicable to all situations of decision-theoretic certainty, risk, and uncertainty, telling us whether or not we should take precautionary measures in the form of choosing the option with the best worst case. We then get the following ranking of the five candidates with respect to their scope:

$$TPA = P\,3 = MaximinPP > UUP > RCPP$$

Simplicity I understand the theoretical virtue of simplicity as demanding that the conceptual apparatus of the target system should be economical in the sense that the concepts it includes that cannot be reduced to each other are kept to a minimum. See also p. 118.

The theoretical apparatus of the Maximin-PP consists of seven concepts: probability, likelihood, outcome, in/commensurable, comparable, course of action, reasonable outcomes (ordinal/cardinal can be reduced to in/commensurable and comparable; decision theoretic risk/uncertainty/partial uncertainty can be reduced to probability, likelihood, and outcome).

The theoretical apparatus of the TPA consists of eight concepts: uncertainty, serious (threat), safety target, knowledge condition, stringency (of knowledge conditions), consistency, efficiency, precautionary measure.

For all five candidates, we get the following ranking with respect to simplicity:

$$UUP\ (5) > RCCP\ (7) = MaximinPP\ (7) > TPA\ (8) > P\,3\ (13)$$

Overall Ranking: Theoretical Virtues of the Five Candidate Systems As an overall ranking with respect to the theoretical virtues, we get the following:

Determinacy:	$UUP = MaximinPP > TPA > RCPP > P3$
Practicability:	$MaximinPP > UUP > TPA > RCPP > P3$
Scope:	$TPA = P\,3 = MaximinPP > UUP > RCPP$
Simplicity:	$UUP\ (5) > RCCP\ (7) = MaximinPP\ (7) > TPA\ (8) > P\,3\ (13)$

The direct comparison in pairs of each candidate system with respect to each virtue is summarized in Table 7.1. There is no overall pareto optimal option, but some partial orderings can be obtained.

The Maximin-PP outranks the RCPP with respect to every virtue, i.e., is pareto optimal compared with the RCPP. Thus, we can note that MaximinPP > RCCP with respect to theoretical virtues.

Also, the Maximin-PP and the TPA have the same scope (range of applicability), and the Maximin-PP otherwise ranks higher than the TPA; the Maximin-PP is therefore the pareto-optimal option and we have Maximin PP > TPA.

As with the TPA, the Maximin-PP is pareto optimal when compared with the P 3-System: Maximin PP > P 3.

The UUP ranks higher than the RCPP in every aspect, so we have UUP > RCPP with respect to all theoretical virtues.

Table 7.1 Detailed comparison of theoretical virtues of candidate systems

		Determinacy	Practicability	Scope	Simplicity
MaximinPP	vs. RCPP	$MPP > RCPP$	$MPP > RCPP$	$MPP > RCPP$	$MPP > RCPP$
	vs. TPA	$MPP > TPA$	$MPP > TPA$	$MPP = TPA$	$MPP > TPA$
	vs. UUP	$UUP = MPP$	$MPP > UUP$	$MPP > UUP$	$UUP > MPP$
	vs. P 3	$MPP > P3$	$MPP > P3$	$P3 = MPP$	$MPP > P3$
RCPP	vs. TPA	$TPA > RCPP$	$TPA > RCPP$	$TPA > RCPP$	$RCCP > TPA$
	vs. UUP	$UUP > RCPP$	$UUP > RCPP$	$UUP > RCPP$	$UUP > RCCP$
	vs. P 3	$RCPP > P3$	$RCPP > P3$	$P3 > RCPP$	$RCCP > P3$
TPA	vs. UUP	$UUP > TPA$	$UUP > TPA$	$TPA > UUP$	$UUP > TPA$
	vs. P 3	$TPA > P3$	$TPA > P3$	$TPA = P3$	$TPA > P3$
UUP	vs. P 3	$UUP > P3$	$UUP > P3$	$P3 > UUP$	$UUP > P3$

The TPA is pareto optimal when compared with the P 3-System, so we get the overall ranking with respect to theoretical virtues: TPA > P3.

Partial Orderings The assessment of the five candidate systems with respect to their theoretical virtues results in the following partial orderings:

$$TPA > P3$$

$$UUP > RCPP$$

$$Maximin\ PP > RCCP$$

$$Maximin\ PP > TPA$$

$$Maximin\ PP > P\ 3$$

What is missing is the overall ranking between Maximin-PP and UUP, between TPA and RCPP, between RCPP and P 3-System, and between UUP and P 3-System.

The trade-offs involved are:

The Maximin-PP is as determinate as the UUP, more practicable and has a higher scope, but is less simple. At this point, it is unclear what the best way is to trade this off. I will come back to it when giving an overall assessment and comparison of candidate systems with respect to theoretical virtues *and* account.

The TPA ranks higher than the RCPP with respect to all virtues aside from simplicity, where the theoretical apparatus of the TPA includes one concept more. I thus argue that we can adopt the following ranking:

$$TPA > RCPP$$

The RCPP ranks lower than the P 3-System with respect to scope, but otherwise ranks higher than it. Again, it is unclear how to best resolve this trade-off, and I will come back to it when giving an overall assessment and comparison of candidate systems with respect to theoretical virtues *and* account.

The UUP has a smaller scope than the P 3-System, but otherwise ranks higher than it. Again, I will consider this when including account in the next subsection. (It makes sense that trade-offs involving scope are difficult to assess without also considering account.)

For now, we can obtain the following partial orderings of the candidate systems with respect to their theoretical virtues:

$$MaximinPP > TPA > RCPP$$

$$MaximinPP > TPA > P3$$

It is still an open question how RCPP and P 3 compare, and where the UUP would belong in this overall ranking. I address these questions in the next section, when coming to the overall ranking of the five candidates with respect to both account and theoretical virtues.

7.4.3 *Overall Comparison of Maximin-PP, TPA, P3, RCPP, and UUP*

With respect to **account**, the five candidates rank as following (compare Fig. 7.7, p. 174):

$$TPA(150) > P3(137) > MaximinPP(127.5) > RCPP(78) > UUP(42)$$

We can already establish that the RCPP is not a candidate that should be considered at this point, since it ranks lower than both Maximin-PP and TPA with respect to overall theoretical virtues and with respect to account.

I also argue that we should eliminate the UUP since its account value is so low that the gains in simplicity cannot outweigh it—especially because the Maximin-PP is on a par with the UUP with respect to determinacy, and better with respect to practicability.

The P 3-System, while reaching the second-highest account value, ranks lower than both the Maximin-PP and the TPA with respect to overall theoretical virtues. Especially its low rankings with respect to determinacy and practicability recommend against adopting it, if we are searching for an action-guiding principle. I thus argue that we should eliminate it at this point, too.

This leaves us with a choice between the Maximin-PP and the TPA. While the TPA has a higher account value than the Maximin-PP, they are not that far apart from each other. And the Maximin-PP ranks higher than the TPA with respect to the overall theoretical virtues. I thus argue that at this point of the RE process, it makes sense to select the Maximin-PP as the current system S_4, and to explore whether its account value can be increased.

7.5 Step B_4: Adjusting Commitments to the Maximin-PP

When adjusting commitments in order to increase account (Sect. 7.5.1), problems for the Maximin-PP are posed especially by commitments that are consistent, but not accounted for by the Maximin-PP. These are mostly

(continued)

moral value commitments, on which the Maximin-PP remains silent. When searching for further relevant commitments (Sect. 7.5.2), more such value commitments emerge. This even slightly decreases the account of the Maximin-PP for the current commitments at the end of step B_4 (Sect. 7.5.3), as Fig. 7.8 summarizes.

7.5.1 Trying to Increase Account

As Fig. 7.7, p. 174, shows, none of the current explicit commitments conflicts with the Maximin-PP, which was adopted as the current system. However, it is still possible to increase account by adjusting commitments, if we can somehow adjust those that the Maximin-PP is consistent with but cannot account for, or respectively can only partially account for. We start with the one that so far is only partially accounted for.

Commitment That Is Partially Accounted for by the Current System I had the following commitment on how the seriousness of threats should be assessed and compared for the application of the target PP:

EC 3 The seriousness of a threat is assessed according to (i) the potential for harm of the threat, and (ii) whether or not the possible harm is seen as reversible. [low] [emerged at Step B_1]

This is only partially accounted for by the Maximin-PP, since while the potential for harm will play a role, irreversibility is not explicitly named. But the Maximin-PP assesses threats according to negative utility and incommensurability, and arguably the relevant sense of irreversibility can be understood along the lines of incommensurability. Thus, I propose to replace EC 3 by the following commitments:

C 5 The seriousness of a threat is assessed according to (i) the potential for harm of the threat, and (ii) whether or not this harm is incommensurable (e.g., because of being irreversible in some relevant sense) with other outcomes. [low] [replaced EC 3 at Step B_4]

Next, I discuss the commitments that are consistent with, but not accounted for by the Maximin-PP.

Commitments That Are Consistent with the Current System, but Not Accounted for by It The first commitment that is consistent but not accounted for concerns one of the toy examples:

IC 11 In case 9, *Job Offers*, you should choose the job in Chicago. [medium]

Given the Maximin-PP, choosing the job in Chicago is not a precautionary measure—it is not the option with the best worst case. Yet the Maximin-PP does not

Step B₄: Current Set of Commitments C₄

General Commitments on Precaution and Precautionary Decision-Making

	C1	C2	IC3	IC4	IC5	IC6	IC7	IC8	EC1	EC2	C5	C6	C3	EC16	EC17	EC18	EC19	EC20	EC21	EC22
P3-System	✓2	-0.5	-1.5	-1	✓6	✓6	-1	✓2	✓2	✓2			✓4	-1.5	✓4	~1	✓2	✓4	✓4	✓4
RCPP	-0.5	-0.5	-1.5	-1	✓6	✓6	-1	✓2	✓2	-0.5			✓4	-1.5	-1	✓2	× -2	× -4	-1	-1
UUP	-0.5	-0.5	-1.5	-1	× -6	-1.5	-1	× -2	✓2	-0.5			× -4	-1.5	-1	× -2	× -2	× -4	× -4	-1
Maximin-PP	-0.5	-0.5	-1.5	-1	✓6	✓6	-1	✓2	✓2	-0.5	✓2	-0.5	✓4	-1.5	-1	✓2	✓2	✓4	✓4	✓4
TPA	✓2	✓2	✓6	-1	✓6	✓6	-1	✓2	✓2	✓2			✓4	-1.5	-1	✓2	✓2	✓4	✓4	✓4

	EC23	EC24	EC25	EC26	EC27
P3-System					
RCPP					
UUP					
Maximin-PP	-0.5	-0.5	-0.5	-1.5	-0.5
TPA					

✓ full account		+2
~ partial account		+1
consistent, but no account		-0.5
× Conflict		-2
black: high weight	* 3	
grey: medium weight	* 2	
white: low weight	* 1	

Commitments on Toy-Examples

	IC9	IC10	IC11	C4	IC13	IC14	IC15	IC16	IC17	IC18	IC19	IC20	IC21	IC22
P3-System	✓6	✓6	✓4	-1	✓6	✓6	✓4	✓4	✓4	✓6	✓4	✓6	✓4	✓6
RCPP	✓6	✓6	-1	-1	✓6	✓6	-1	✓4	✓4	✓6	✓4	✓6	✓4	✓6
UUP	✓6	✓6	-1	-1	✓6	✓6	✓4	✓4	-1	✓6	✓4	× -6	× -4	✓6
Maximin-PP	✓6	✓6	✓4	-1	✓6	✓6	✓4	✓4	✓4	✓6	✓4	✓6	✓4	✓6
TPA	✓6	✓6	✓4	-1	✓6	✓6	✓4	✓4	✓4	✓6	✓4	✓6	✓4	✓6

Commitments concerning Precaution, Climate Change, and Solar Radiation Management (SRM)

	IC23	IC24	IC25	IC26	IC27	IC28	IC29	IC30	IC31
P3-System	✓6	~2	✓4	✓4	✓4	✓6	✓4	-1	✓6
RCPP	✓6	~2	-1	✓4	✓4	-1.5	-1	-1	✓6
UUP	✓6	~2	✓4	✓4	✓4	✓6	✓4	✓4	✓6
Maximin-PP	✓6	~2	✓4	✓4	✓4	✓6	✓4	-1	✓6
TPA	✓6	~2	✓4	✓4	✓4	✓6	✓4	✓4	✓6

Fig. 7.8 End of Step B_4: current commitments C_4

recommend against it, either. So actually, being consistent with this commitment might be enough. For now, I keep it like this in the set of relevant commitments, but we can mark it as a candidate for adjustment.

EC 2 Serious threats *pro tanto* warrant precaution. [low] [emerged at Step B_1]

If we accept the Maximin-PP, then it makes sense to say that it picks out those threats that *all things considered* warrant precaution. However, it does not express any *pro tanto* claims and thus cannot account for this commitment. Should we reject the commitment? I do not think that this is defensible: that serious threats *pro tanto* warrant precaution seems almost trivially true. However, it might not be necessary that the target system can account for it, so maybe it will make sense to exclude it from the subject matter. But before making this decision, I will first compare further candidate systems.

EC 4 All *plausible* serious threats *pro tanto* warrant precaution. [low] [emerged at Step B_1]

If we think that all plausible threats are somewhere between uncertainty and risk, then the Maximin-PP would be able to account for this. I propose the following adjustment:

C 6 *Pro tanto*, every serious threat warrants precaution as long as it meets some minimal criteria of plausibility or reasonableness—i.e., that the likelihood of a possibility of severe harm is very low or cannot even be assigned is no *pro tanto* reason against taking precautions. [high] [replaced EC 4 at Step B_4]

The Maximin-PP can account for C 6 because every threat that meets at least some minimal criteria of plausibility that make it reasonable enough to be included in the outcome-set of a course of action will play a role for deciding which course of action to take.

The next commitment that is consistent with the Maximin-PP but not accounted for concerns the distribution of costs and responsibilities for precautionary measures:

EC 16 The costs and responsibilities for precautionary measures should be distributed in a morally sound way. [high] [emerged at Step B_2]

Here, we could argue that maybe such matters are outside of the subject matter—that it is enough if our target system can identify which measures are recommended or demanded by moral precaution. Still, at the moment this argument seems too weak to fulfill the respecting-condition, given that the commitment has a high weight. EC 16 remains in the current commitments.

There are also three commitments that already were adjusted, but are not accounted for by the Maximin-PP. First, we have C 4:

C 4 The claim that the measles/mumps/rubella (MMR) vaccine might cause autism does not constitute a plausible threat. [medium] [replaced IC 12 at Step B_3]

If we understand this as an epistemic claim, then we might have a good enough reason to exclude it from the subject matter, since the pragmatic-epistemic objective is to identify and defend a *moral* principle. But at this point, I am undecided. I keep it and re-assess it in the next step.

The remaining two consistently non-accounted for commitments are C 1 and C 2:

C 1 When there is a plausible threat of serious or irreversible harm to the environment, then uncertainty of the harm must not lead to postponing cost-effective measures to prevent it. [medium] [replaced IC 1 at Step B_3]

C 2 When an activity raises threats of harm to human health or the environment, then, *pro tanto*, precautionary measures should be taken even if some cause-and-effect relationships are not fully established scientifically. [medium] [replaced IC 2 at Step B_3]

These two commitments express claims about how decisions should be made and translated into actions (C 1), or respectively demanding that actions have to be taken when there are specific kinds of threats (C 2). Both of them are consistent with the Maximin-PP: we could, e.g., use the Maximin-PP to come to decisions in accord with C 1, and we can interpret harms to human health or the environment as incommensurable with other kinds of outcomes, thereby also supporting the Maximin-PP as a decision-principle that could be used in these cases. But neither of the two is implied by the Maximin-PP itself, i.e., none of them is accounted for by the current System S_4.

But this, rather, gives support to the idea that the current system should be adjusted in some way, and not that C 1 or C 2 should be rejected or otherwise adjusted.

Thus, out of a broad range of commitments that are consistent with, but not accounted by, the Maximin-PP, only one could be defensibly adjusted at this point, namely EC 4 which was replaced by C 6.

7.5.2 Searching for Further Relevant Commitments

While S_4, the Maximin-PP, cannot account for all of the commitments, most of them are accounted for by it and the remaining few are at least consistent with it. But if we look at these "remaining few", then the concern arises that they might not actually express something important and central to the subject matter—and for our pragmatic-epistemic objective, too: if we want the target PP not only to be *normative*—e.g., about rationality—but really a *moral* principle, then it seems that there are further relevant, substantial commitments that should be accounted for.

I already have several precautions concerning human health and the environment, e.g.

IC 1 Where there are threats of serious or irreversible damage, lack of full scientific certainty shall not be used as a reason for postponing cost-effective measures to prevent environmental degradation. (Principle 15 of the Rio Declaration) [low]

IC 2 When an activity raises threats of harm to human health or the environment, precautionary measures should be taken even if some cause-and-effect relationships are not fully established scientifically. (Wingspread Formulation of the Precautionary Principle) [low]

IC 4 Threats to the environment or to human health warrant special precaution because such harm is especially prone to have long latent periods, and to be hard if not impossible to remediate or compensate. [medium]

IC 3 *Pro tanto*, it is better to take precautionary measures now than to deal with serious harms to the environment or human health later on. [high]

I take it that protection of human health and the environment is a core claim of the target PP, which warrants making the following two commitments explicit:

EC 23 *Pro tanto*, threats of harm to human health have lexical priority for precaution. [low] [emerged at Step B_4]

EC 24 *Pro tanto*, threats to the environment have lexical priority for precaution. [low] [emerged at Step B_4]

Then there are distributional concerns, which also matter for precaution (cf. Sunstein 2007, 2). I am committed to a precautionary principle as standing against myopic decisions that unfairly advantage the present at the cost of the future by avoiding efficient action to address threats while there is still time. Also, when precaution is seen as concerned with worst cases, I understand this not only as the worst *net* outcome, but also as a requirement to pay special attention to those who would be worst off (cf. Bognar 2011, 341–42).

EC 25 When evaluating possible outcomes of courses of actions, the rights of future generations must not be discounted. [low] [emerged at Step B_4]

EC 26 When taking precautionary measures against a threat, attention has to be paid to those who would be worst off if the harm should materialize. (Distributional concerns matter for precaution.) [high] [emerged at Step B_4]

EC 27 Serious threats that can be addressed by an earlier generation must not be deferred to future generations. [low] [emerged at Step B_4]

7.5.3 *The Adjusted Set of Current Commitments, C_4*

The current set of commitments, C_4, and its agreement with the current system—the Maximin-PP—is summarized in Fig. 7.8. By adjusting EC 3 and EC 4 to C 5 and C 6, the account value could be increased. However, there are also the five emerging commitments EC 23–EC 27 that the Maximin-PP cannot account for and is merely consistent with, which again decreases its account value. All in all, the Maximin-PP reaches an account value of 126.5 at the end of step B_4, which indeed is a small decrease from 127.5, which it reached when accounting for the commitments C_3.

7.6 Recapitulation Phase 2

In Fig. 7.9, the results of the steps of phase 2 are summarized. We started by comparing the Principle 3-System, which was formulated at the end of phase 1, with the Rawlsian Core Precautionary Principle (RCPP) and the Utilitarian Uncertainty Principle (UUP). After selecting the RCPP based on its theoretical virtues and its account for commitments, we moved on to adjust the current commitments from C_2 to C_3, which included replacing IC 1, IC 2, EC 15, and IC 12 by C 1–4, and making explicit two further emerging commitments, EC 21 and EC 22. In the next step, A_4, two new candidate systems were introduced, the Maximin-Precautionary Principle for Combinations of Uncertainty and Incommensurability (Maximin-PP), which is a result of adjusting the RCPP, and the Tripartite Precautionary Approach (TPA). The Maximin-PP was selected as the most defensible candidate, and the commitments were adjusted with respect to it from C_3 to C_4 at step B_4. This consisted of replacing EC 3 and EC 4 by C 5 and C 6, and in making further commitments explicit that mostly are commitments to values and evaluations, EC 23–EC 27.

7.6.1 Phase 2: Discussion of Intermediate Results for RE

Main results from phase 2 for reflective equilibrium are:

- The RE criteria put real constraints on the justification-process;
- Whole classes or subsets of commitments can be adjusted;
- Re-interpreting commitments is also a way of adjusting them;
- The RE criteria can be used to improve candidate systems.

As an important result of phase 2, we can note that the RE criteria put real **constraints on the justification process**: in step A_3, the P 3-System is rejected because of its low systematicity, whereas the UUP is rejected because of its low account value, which cannot be outweighed by its high systematicity. In both cases, the costs in terms of negative effects on the position are too high compared with the benefits of the two candidates. While these decisions are not absolute and incontestable, they are not arbitrary, either. In principle, one could try, e.g., to accept the UUP and explore whether in the long run one arrives at a more plausible position than when accepting the RCPP at this step—maybe by adjusting some of the commitments, but also supplementing the UUP with other parts of the system. But given the evaluation of all three candidates, it is defensible to argue that the RCPP is the most convincing candidate at the end of step A_3.

As step B_3 shows, adjusting commitments does not only concern individual commitments: we can also **consider whole classes or subsets of commitments**. In

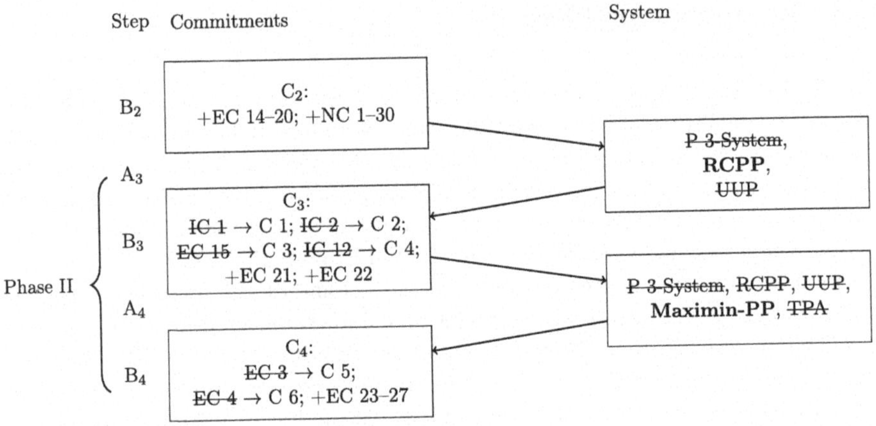

Fig. 7.9 Schematic overview of the Steps of Phase 2

this step, I considered whether certain subsets of commitments should be excluded from the subject matter of precautionary principles, which would be a form of rejecting them from the current justificatory project. That is, I did not investigate particular commitments, but asked whether, in general, commitments on cases where probabilities are available or commitments on cases without or with more than one acceptable worst case should be rejected as irrelevant for a precautionary principle. This was prompted by the fact that the best PP candidate at this stage, the RCPP, cannot account for commitments on such cases. Consequently, it would increase the RCPP's account value to reject these subsets. I argued that these subsets of commitments are, indeed, important parts of the subject matter of (moral) precaution and precautionary decision-making. For example, the possibilities of disproportionate and/or asymmetric harm (see Fig. 7.4c) lend support to the idea that a PP should be applicable at least in some cases where probabilities are available. While in this case, the subsets survived their challenge from the current candidate system, this illustrates that if no such independent arguments were available, it would be possible to reject or adjust whole subsets of commitments.

Another result concerning the adjustment of commitments is illustrated by C 1 and C 4 which are **re-interpretations of commitments**. They are not adjustments in the sense of replacing a commitment that conflicts with the system, but in the sense of clarifying vague commitments. This is similar to explications, where a pre-theoretical concept gets replaced by a new concept that is specifically constructed for a theoretical purpose: in our input commitments, we might have vague commitments such as IC 12, where it is unclear what "not being a reason" is supposed to express, exactly:

IC 12 The claim that the measles/mumps/rubella (MMR) vaccine might cause autism is not a reason not to vaccinate your child. [medium]

In the RE process, we can then clarify such commitments with respect to the current system—whose selection was based on the whole set of commitments. In step B3, this meant replacing IC 12 with C 4:

C 4 The claim that the measles/mumps/rubella (MMR) vaccine might cause autism does not constitute a plausible threat. [medium] [replaced IC 12 at Step B3]

When moving from the RCPP to the Maximin-PP, it was demonstrated how the RE criteria can help to **improve candidate systems**: weaknesses of the RCPP were identified as results of steps *A*3 and B3, and the Maximin-PP is an attempt at keeping the positive properties of the RCPP but improving its weaknesses.

In step B4, further emerging commitments are made explicit that are mostly value-commitments. Those commitments are relevant for the subject matter, but were not explicitly considered before, because other problems had to be tackled first. Substantial value commitments only came into focus after some structural problems were at least tentatively addressed—by formulating a candidate system that names conditions for when a specific course of action should be chosen. This shows again that **RE often proceeds piecemeal**, even if it is ultimately a holistic process.

7.6.2 Phase 2: Discussion of Intermediate Results for PPs

Main results from phase 2 for the justification of a moral precautionary principle are:

- Precaution is not about maximizing expected utility, but concerns specific threats of harm that cannot be outweighed by potential benefits;
- The explication of "being a precautionary measure against an undesirable *x* (*ExplicPrec*)" moved to the background.

Firstly, the results from phase 2 can be used to reject the claim by Bognar (2011, 339) that "Whenever the Rawlsian conditions are approximated, the core precautionary principle offers no advantage over utilitarianism", i.e., that the RCPP should be replaced by the UUP since the latter will yield the same verdicts as the RCPP whenever the RCPP yields a verdict, but is also rational in other situations where the conditions of the RCPP are not met. There are two main reasons to reject this claim, which can be made on the basis of the results of the RE process: (1) The RCPP can account for some important commitments about precaution, where the UUP fails to account for them or even leads to conflicts. Examples are IC 5 or IC 6. (2) There are situations where the conditions of the RCPP are met, and the verdict of the UUP conflicts with the one of the RCPP: the RCPP can account for the case-specific commitments IC 20 and IC 21, whereas the UUP conflicts with them.

This suggests that precaution is not about maximizing expected utility, but focuses on specific threats of harm that cannot be outweighed by potential benefits. That harms cannot be outweighed by benefits here does not simply mean that in direct comparison the harm is worse than the benefits, but also refers to situations in which there are, e.g., no second chances: accepting a threat that could lead to the destruction of the earth cannot be outweighed by any chance of an extremely good paradise-on-earth outcome (at least if the two are equally plausible).

Secondly, we can note that the explication of "being a precautionary measure against an undesirable *x* (*ExplicPrec*)" lost its relevance. Both the RCPP and the Maximin-PP will only select measures that fulfill the criteria for being a precautionary measure, without needing *ExplicPrec* to yield a verdict. They thereby meet the desideratum that was formulated at the end of Sect. 6.8.2, namely to identify "a system that can pick out justified cases of precautionary measures *without* having to refer to such an additional explication". This does not mean that *ExplicPrec* becomes completely superfluous, but we can exclude it from the current position in the foreground and move it to the background. There, it fulfills the role of making sure that the target system will not lead to verdicts that can't be characterized as *precautionary*.

The last point is not a result, but rather a caveat: all the assessments and results so far depend on the stipulation of a "reasonable outcomes" criterion (see Sect. 6.4.2). This means that assessment of account, and thus the selection of the current system, is contingent on whether or not such a criterion can be identified.

References

Aldred J (2013) Justifying precautionary policies: incommensurability and uncertainty. Ecol Econ 96:132–140. https://doi.org/10.1016/j.ecolecon.2013.10.006

Bognar G (2011) Can the maximin principle serve as a basis for climate change policy? Monist 94(3):329–348. https://doi.org/10.5840/monist201194317

Gardiner SM (2006) A core precautionary principle. J Polit Philos 14(1):33–60

Hansson SO (2008) From the casino to the jungle. Synthese 168(3):423–432. https://doi.org/10.1007/s11229-008-9444-1

Knight C (2017) Reflective equilibrium. In: Blau A (ed) Methods in analytical political theory. Cambridge University Press, Cambridge, pp 46–64

Randall A (2011) Risk and precaution. Cambridge University Press, New York

Roser D (2017) The irrelevance of the risk-uncertainty distinction. Sci Eng Ethics 23(5):1387–1407. https://doi.org/10.1007/s11948-017-9919-x

Steel D (2015) Philosophy and the precautionary principle. Cambridge University Press, Cambridge

Sunstein CR (2007) The catastrophic harm precautionary principle. Issues in Legal Scholarship 6(3):1–29

Chapter 8
Case Study, Phase III: Reaching a State of Reflective Equilibrium?

In the third and final phase of the case study, I work toward a preliminary consolidation and evaluation of a resulting position. Previously, in the first phase (Chap. 6), I tested how reflective equilibrium (RE) can be used to construct a first candidate system, and in the second phase (Chap. 7) we saw how the RE criteria can be used in the two alternating steps of adjusting system and commitments. Now the case study focuses on the (preliminary) conclusion of the equilibrium-process.

8.1 Overview: Phase 3

In this final phase of the case study, the goal is to test how the RE criteria can be used to assess a resulting position, i.e., to assess whether it is in a state of reflective equilibrium. Thus, to reach a position that is sufficiently fleshed out, the selection of alternative candidate systems is narrowed down for the purpose of the case study.

In Sect. 8.2, we start by adjusting the current system—the Maximin Precautionary Principle for Combinations of Uncertainty and Incommensurability (Maximin-PP)—in order to enable it to account for the commitments that remained unaccounted for at the end of Step B_4 (see Chap. 7). Several candidates for a substantial moral-value base of a PP are compared, but only one candidate—a rights-based approach to moral precaution—is further explored and elaborated.

When adjusting the current commitments with respect to the chosen "Rights-Maximin-PP" in Step B_5 (Sect. 8.3), further input commitments emerge that are in tension with the Rights-Maximin-PP. When adjusting the system in Step A_6 (Sect. 8.4), I thus compare the Rights-Maximin-PP with another new candidate, the Tripartite Precautionary Approach to Threats of Rights Violations (Rights-TPA), which turns out to be more defensible than the Maximin-PP, and is selected as the resulting system at the end of phase 3. Arguably, in Steps A_7 and B_7, applying the two steps of adjusting system and commitments no longer leads to

T. Rechnitzer, *Applying Reflective Equilibrium*, Logic, Argumentation & Reasoning 27, https://doi.org/10.1007/978-3-031-04333-8_8

substantial changes of the position. Consequently, the equilibration-process comes to a (preliminary) end point, and, in Sect. 8.6, I evaluate whether, and to what degree, a position in reflective equilibrium was reached.

Section 8.7 recapitulates the results from phase 3, including a schematic summary of the steps in Fig. 8.5. In the appendix, you can also find Fig. A.7, which gives a schematic overview of the whole process of adjustments.

Throughout phase 3, gray boxes are again used to summarize the main points of each step. As before, only relevant or exemplary aspects of the process are described in detail, and readers can refer to Appendix A at the end of the book for the full list of commitments, candidates for (parts of) the system, background information, and case descriptions.

8.2 Step A_5: Developing and Adopting the Rights-Maximin-PP

The Maximin Precautionary Principle for Combinations of Uncertainty and Incommensurability (Maximin-PP) is adjusted in order to be better able to account for commitments. The meaning of "incommensurability" in the Maximin-PP is newly explicated with "(threshold) lexical superiority", i.e., it applies when some outcome values cannot be outweighed by others because they take lexical priority (Sect. 8.2.1). As candidates for a relevant threshold of lexical priority, human rights, environmental harm, irreversible harm, harm to human health, and catastrophic harm are roughly assessed. To obtain a new candidate system, I then propose to supplement the Maximin-PP with *The Rights-Threshold Principle*, i.e., giving lexical priority to avoiding wrongful rights violations (Sect. 8.2.2). After assessing this *Rights-Maximin-PP* with respect to its ability to account for current commitments (Sect. 8.2.3) and its theoretical virtues (Sect. 8.2.4), I adopt it as the new system (Sect. 8.2.5).

In Step A_4 (Chap. 7), the Maximin-PP was chosen as the current system:

Maximin-PP for Combinations of Uncertainty and Incommensurability (Maximin-PP) Select the course of action with the best worst case if you are either:

- In a situation of decision-theoretic risk or uncertainty (or some combination), and the outcomes of the available actions can be ranked on an ordinal scale, and all courses of action alternative to the one selected by maximin have outcomes that are incommensurably worse than the best worst case; or

- In a situation of (partial) decision-theoretic uncertainty, outcomes can be ranked on a cardinal scale, and all courses of action alternative to the one selected by maximin have negative outcomes that outweigh every potential gain that could be made above the level that can be guaranteed by maximin.

The Maximin-PP tells us that if there are threats of harm that for some reason or another cannot be outweighed by possible gains, then we should choose the course of action that has a best worst case that does not threaten to cause this kind of harm. However, as we have seen when adjusting the commitments in Step B_4 (Chap. 7), the Maximin-PP cannot, in itself, account for a range of substantial value commitments that concern, e.g., the protection of human health, the environment, or the rights of future generations. The Maximin-PP is a principle of *rational choice* that gets applied to a decision problem in which we already know the values that we assign to the various possible outcomes. However, the pragmatic-epistemic objective of my RE project is to justify a principle of *moral* precaution (see Chap. 5). I am committed to a difference between rational, self-interested precaution, and morally demanded precaution. When exposing yourself to an uncertain harm, this is a question of rationality.[1] However, when you expose others to uncertain outcomes, the demands of morality additionally come into play. There is a difference between risk-taking and risk-imposing.

IC 7 Morally, a higher degree of precaution is required when making decisions that will have effects on others: when making decisions that will only affect yourself, precaution is a question of rationality, depending on your preferences and beliefs; but when making decisions that threaten to harm others, precaution is morally required. [medium]

This is one of the commitments that the Maximin-PP cannot account for. We could now discard the Maximin-PP, and try to come up with a completely new candidate system. Instead, though, I am going to try to adjust the Maximin-PP and to develop it into a moral precautionary principle. Such a substantial moral precautionary principle does not need to conflict with what rationality requires, but it might put additional requirements on our decisions. In Step B_4, a range of input commitments emerged that assign more weight, or even lexical priority, to certain kinds of threat:

EC 23 *Pro tanto*, threats of harm to human health have lexical priority for precaution. [low] [emerged at Step B_4]

EC 24 *Pro tanto*, threats to the environment have lexical priority for precaution. [low] [emerged at Step B_4]

EC 26 When taking precautionary measures against a threat, attention has to be paid to those who would be worst off if the harm should materialize. (Distributional concerns matter for precaution.) [high] [emerged at Step B_4]

---

[1] Assuming that only you are affected, and that there are, e.g., no indirect effects on people who care about you, etc.

Thus, the moral precautionary principle that we are searching for might in particular put additional requirements on how possible outcomes should be evaluated. What form can, or should, these additional requirements take? In the following, I propose (i) to use another explication of "incommensurability", i.e., to spell it out in terms of lexical priority, and (ii) to supplement the Maximin-PP with a threshold that gives lexical priority to human rights.

8.2.1 Explicating "Incommensurable" as "(Threshold) Lexical Superiority"

So far, I did follow Aldred (2013, 133) in defining incommensurability of outcomes as meaning that their value cannot be precisely measured along some common cardinal scale. However, that the values of two outcomes are incommensurable does not seem to be enough to warrant choosing the course of action with the best worst case as the Maximin-PP demands. Incommensurability as defined by Aldred only entails that we do not know, e.g., how much better or worse one outcome is than another. It does not entail that some outcomes are *always* better or worse than other outcomes—i.e., that there are some values of outcomes that take lexical priority (cf. Chang 2013). While incommensurability in the sense of values not being measurable along a common cardinal scale is part of (threshold) lexical priority, it is not already sufficient to establish it. Yet lexical priority seems to be what should be required for the Maximin-PP: that some outcome values are always worse or better than any instance of other outcome values (Chang 2013, 3–4). This understanding of "incommensurability" also fits better with Aldred's own example, in which the medium outcome (reduced economic growth from climate change mitigation) is always better than the worst case (climate catastrophe):

> The key discontinuity claim is that, no matter how much worse we make *m* (call it *m*− − −), it is still better than *w*. *m*− − − involves very high mitigation expenditure, but it is still better than any outcome *w* involving climate change catastrophe. *w* is incommensurably worse than both *b* [no climate catastrophe, no mitigation costs, *T.R.*] and *m* (which are commensurable with each other). (Aldred 2013, 137)

The Maximin-PP tells us to choose the policy option that has *m* as its possible outcome, and not the one that has *w* and *b* as its possible outcomes. Now, especially the claim that "no matter how much worse we make *m* [...], it is still better than *w*", indicates that there is more at stake than outcome values not being measurable among a common cardinal scale: moreover, avoiding some outcome value (climate catastrophe) takes *lexical priority* over promoting other outcome values (additional economic gains).[2]

[2] Both in the "basic" sense of incommensurability as well as with lexical priority, outcome values might still be *comparable*, i.e., it is not excluded that they can be ranked on an ordinal scale.

Consequently, I propose to adjust the meaning of "incommensurability" in the Maximin-PP to refer to "lexical priority". The next question is which threshold(s) of lexical priority should be chosen to supplement the Maximin-PP in order to enable it to account for the substantial value commitments that, so far, it cannot account for.

8.2.2 *Candidates for a Threshold of Lexical Priority*

There are several candidates for having lexical priority when it comes to taking precautionary measures. Among the most prominent that we can find in the literature are: harm to the environment, harm to human health, irreversible harm (these three can, e.g., be found in both the Rio and the Wingspread PP), catastrophic harm (e.g., Hartzell-Nichols 2012, 2017; Sunstein 2007), and violations of rights (Caney 2009; Roser 2009, 2020).

Human Rights I argue that based on the subject matter, i.e., my commitments, and my pragmatic-epistemic objective, human rights are a good candidate for having lexical priority when it comes to taking precautionary measures. Firstly, rights are already seen as constituting such a threshold: "Rights are characterized by a threshold—not letting other persons fall below that threshold is of very high (or absolute) importance, benefiting them above the threshold is of very low (or zero) importance" (Roser 2009, 16).

Secondly, I argue that adopting rights as the normative basis for a precautionary principle provides a unifying rationale, since most if not all relevant cases of harm to the environment and/or to human health will be subsumable under it—as will be cases of threat of catastrophe.

In the following, I discuss each of the other candidates in comparison with the rights threshold, arguing that on their own, they all face significant problems and/or can relatively straightforwardly be subsumed under a rights threshold.

Environmental Harm Harm to the environment does not only raise conceptual questions such as how to distinguish "nature" from "culture": there is also the fundamental question of *why* we should give priority to avoiding harms to the environment. Is it because we ascribe some intrinsic value to the environment? But if yes, does this value have lexical priority compared with basic human interests?

I am not willing to commit to, e.g., that we should have let Hurricane Katrina run its course, as Hartzell-Nichols (2013) suggests would have been a consequence of a PP that gives lexical priority to protecting the environment:

> It arguably would have been much better for the environment to let Hurricane Katrina run its course, as reinforcing levees, while important to the protection of human health and property, only further interfered with natural sediment transfer. (Hartzell-Nichols 2013, 313)

I do not want to take a stance here on the question of whether or not our environmental ethics should be anthropocentric, e.g., whether or not we should ascribe value to the environment only insofar as it has instrumental value for human interests. Giving lexical priority to (the protection of) rights as the normative basis of a moral PP does not exclude the possibility of ascribing intrinsic value to the environment. It just means that when there is a conflict between threats of environmental harm and threats of rights violations, the threats to the latter take priority. And since an intact environment is important for even the most basic and fundamental rights, it is to be expected that such conflicts will only seldom or only temporarily (e.g., in case of impending harm, like the Hurricane Katrina example) lead to environmental degradation.

Irreversible Harm Giving lexical priority to the avoidance of irreversible harm is not defensible either: firstly, there is the question of how to conceptualize the relevant sense of "irreversible", since it cannot mean everything that cannot be undone, like the decision to take coffee instead of tea for breakfast in the hotel (which could even cause me some small irreversible *harm* if the coffee turns out to be disgusting). Secondly, even if there is a plausible way to conceptualize irreversibility, it seems rather to be something that reinforces the demand for precaution instead of constituting it on its own. Most importantly, there are threats that demand precaution even if the harm is not irreversible in the relevant sense. And even if it is irreversible, this is not always a reason for extra precaution: at least some goods can be replaced by substitutes that serve the same purpose at least equally well. And in many cases, there are straightforward reasons to even accept irreversible loss of valuable goods for which there is no substitute, as Roser (2020) argues:

> [There] are many straightforwardly justifiable reasons for irreversibly giving up goods. Irreversible loss is a common occurrence on which there is no absolute prohibition. Heritage conservation does not protect every building or valuable memory. Thus, cautiousness with respect to irreversibly lost values needs further argument. (Roser 2020, 309)

In economics, irreversibility is also explicated with the concept of "quasi-option value", i.e., adding an additional positive value to courses of action that "keep options open" by, e.g., not developing/using a natural resource or not permanently polluting something (Arrow and Fisher 1974). Understood in this way, irreversibility can also be relevant from a rights-perspective, additional to the economic argument: in this specific sense, irreversibility can be understood as the opposite of sustainability, and sustainability can, again, be understood as a commitment to the rights of future generations.

Harm to Human Health Harm to human health can most straightforwardly be subsumed under a human rights approach. Not every harm to human health might constitute a human rights violation, but arguably, all the *relevant cases* for precaution will fall under a human rights approach.

Catastrophic Harm As Roser (2020) argues, if "catastrophic" or "serious" harm is about the extent of harm that is threatened, then singling out such thresholds of harm seems *ad hoc*. Since extent of harm is gradual, the response should be gradual, too: of course, serious damage and catastrophes are reason for concern. Any damage is reason for concern and in so far as serious damage amounts to extremely large damage it is reason for extremely large concern. However, the extent of damage is a continuous quantity and—if the focus of the effect condition is put on damage—then there should thus be continuity in the strength of the response as well, rather than a principled difference between the response to serious and non-serious damage.

> Why would a cautious response to uncertainty regarding small damages not be just as appropriate as a cautious response to uncertainty regarding serious damages? Treating serious or catastrophic damage in a fundamentally different way might lead us astray—to take just one example—in comparisons of policies of which one comes with a small probability of catastrophe but is most probably hugely beneficial and another policy has an even smaller probability of catastrophe but virtually certainly yields significant but not quite catastrophic damage. Some reason would have to be given why there should be a non-continuous treatment of damages as they get larger and larger and then cross the threshold to where they are 'catastrophic' or 'serious'. Otherwise, the suggested rationale is *ad hoc*. (Roser 2020, 308)

Catastrophic (or serious) harm is thus not a plausible candidate for having lexical priority as part of the current system, the Maximin-PP. However, harm that threatens to be catastrophic or very serious will typically also threaten substantial rights violations.

Proposing the Rights-Maximin-PP All in all, I argue that giving lexical priority to avoiding rights-violations is the most defensible current alternative for a normative threshold for the evaluation of outcomes. We can formulate the candidate like this:

The Rights-Threshold Principle Threats of rights violations have lexical priority over other threats, and are incommensurable with chances of other kinds of gains.

By combining it with the Maximin-PP, we obtain what I call the "Rights-Maximin Precautionary Principle for Combinations of Uncertainty and Incommensurability (Rights-Maximin-PP)". It is worth pointing out that giving lexical priority to avoid threats of rights violations does not mean that other kinds of threats should be neglected, in the sense that no precautionary measures should be taken against, e.g., threats to human well-being that do not amount to violations of human rights. The point expressed by the Rights-Threshold Principle is that rights deserve special attention: we have to avoid violations of rights even at high costs, as long as these costs do not themselves include equally or more serious rights violations.

In the following, I assess this candidate with respect to its ability to account for commitments and its theoretical virtues. I roughly compare it with the Maximin-PP on its own.

Step A5: Account for Commitments C4

General Commitments on Precaution and Precautionary Decision-Making

	C1	C2	IC3	IC4	IC5	IC6	IC7	IC8	EC1	EC2
Maximin-PP	-0.5	-0.5	-1.5	-1	✓6	✓6	-1	✓2	✓2	-0.5
TPA	✓2	✓2	✓6	-1	✓6	✓6	-1	✓2	✓2	✓2
Rights-Max.-PP	-0.5	-0.5	-1.5	-1	✓6	✓6	✓4	✓2	×-2	-0.5

	C5	C6	C3	EC16	EC17	EC18	EC19	EC20	EC21	EC22
Maximin-PP	✓2	-0.5	✓4	-1.5	-1	✓2	✓2	✓4	✓4	✓4
TPA	-0.5	✓2	✓4	-1.5	-1	✓2	✓2	✓4	✓4	✓4
Rights-Max.-PP	✓2	-0.5	✓4	-1.5	-1	✓2	✓2	✓4	✓4	✓4

	EC23	EC24	EC25	EC26	EC27
Maximin-PP	-0.5	-0.5	-0.5	-1.5	-0.5
TPA	-0.5	-0.5	-0.5	-1.5	-0.5
Rights-Max.-PP	-0.5	-0.5	✓2	✓6	✓2

Commitments on Toy-Examples

	IC9	IC10	IC11	C4	IC13	IC14	IC15	IC16	IC17	IC18	IC19	IC20	IC21	IC22
Maximin-PP	✓6	✓6	✓4	-1	✓6	✓6	✓4	✓4	✓4	✓6	✓4	✓6	✓4	✓6
TPA	✓6	✓6	✓4	-1	✓6	✓6	✓4	✓4	✓4	✓6	✓4	✓6	✓4	✓6
Rights-Max.-PP	✓6	✓6	✓4	-1	✓6	✓6	✓4	✓4	✓4	✓6	✓4	✓6	✓4	✓6

Commitments concerning Precaution, Climate Change, and Solar Radiation Management (SRM)

	IC23	IC24	IC25	IC26	IC27	IC28	IC29	IC30	IC31
Maximin-PP	✓6	~2	✓4	✓4	✓4	✓6	✓4	-1	✓6
TPA	✓6	~2	✓4	✓4	✓4	✓6	✓4	✓4	✓6
Rights-Max.-PP	✓6	✓4	✓4	✓4	✓4	✓6	✓4	-1	✓6

✓ full account +2
~ partial account +1
consistent, but no account -0.5
× Conflict -2

black: high weight * 3
grey: medium weight * 2
white: low weight * 1

Account-Values:

Maximin-PP: 126.5
TPA: 146.5
Rights-Maximin-PP: 143.5

Fig. 8.1 Step A5: account for commitments C4

8.2.3 *Rights-Maximin-PP, Account for Commitments*

Compared with the Maximin-PP without the Rights-Threshold, account was increased from 126.5 to 143.5. See Fig. 8.1 for an overview.

The Rights-Maximin-PP can now account for some of the commitments that the Maximin-PP on its own could not account for:

IC 7 Morally, a higher degree of precaution is required when making decisions that will have effects on others: when making decisions that will only affect yourself, precaution is a question of rationality, depending on your preferences and beliefs; but when making decisions that threaten to harm others, precaution is morally required. [medium]

The Rights-Threshold Principle can account for the difference between risk-taking and risk-imposing: you can waive your own rights, but not those of others.[3]

EC 25 When evaluating possible outcomes of courses of actions, the rights of future generations must not be discounted. [low] [emerged at Step B_4]

If we assume that if you have a right to *x*, then you have this right independently of your place in time and other morally irrelevant factors, then rights of future generations cannot be discounted *simply* because they are in the future.

EC 26 When taking precautionary measures against a threat, attention has to be paid to those who would be worst off if the harm should materialize. (Distributional concerns matter for precaution.) [high] [emerged at Step B_4]

Ensuring that as many people as possible receive what they have a right to takes priority over maximizing net gain (and giving some people more than what they have a right to at the cost of depriving others of their rights).

EC 27 Serious threats that can be addressed by an earlier generation must not be deferred to future generations. [low] [emerged at Step B_4]

If "serious threats" refers to threats to rights, then the current system can also account for this. Maybe the commitment needs to be adjusted—or this counts only as a "partial" account, leaving open the possibility that there might be other classes of serious threat that should not be deferred to future generations.

8.2.4 Rights-Maximin-PP, Theoretical Virtues

I roughly assess the theoretical virtues of the Rights-Maxmin-PP, with respect to the Maximin-PP on its own, and also compared with some of the other candidates for a threshold of lexical priority. For more information on how I understand the theoretical virtues, see Chap. 5 and Sect. 5.5.

Determinacy On the one hand, "incommensurable" was further specified to mean cases of outcome values that have lexical priority over other outcome values. This

[3] I take this to be part of the background, even though there is a debate about whether or not you can actually waive your own fundamental human rights. But in any case, there seems to be an agreement that you can at least waive *some* of your rights to *some degree*.

increases determinacy. On the other hand, the determinacy of the Rights-Threshold Principle depends on how fleshed-out a rights theory we have. Although referring to rights is not extremely determinate, nevertheless the Maximin-PP on its own did not determine any relevant cases of incommensurability, so adding such a threshold does increase its determinacy *even if* this threshold itself is only moderately determinate. And as the comparison of the rights threshold with alternatives like a catastrophic-harm threshold has shown, it is at least as determinate as currently available alternatives.

Practicability As with Determinacy, I argue that the Practicability of the Rights-Maximin-PP is not decreased as compared with the Maximin-PP, since we did spell out one aspect that was not covered by the Maximin-PP and kept the original principle. And compared with other alternatives for having lexical priority, the rights threshold is at least as practicable as them.

Scope Combining the Rights-Threshold Principle with the Maximin-PP does not mean that precaution is *reduced* to threats of rights violations—the Maximin-PP leaves room for other cases of lexical priority and incommensurability, and also still applies to cases where outcomes are commensurable, but disproportional. I have just added the rights threshold as one substantial moral rationale to the Maximin-PP in order to do justice to my pragmatic-epistemic objective of formulating an action-guiding moral precautionary principle that applies in other-regarding decision-making (e.g., intergenerational contexts). I.e., the scope (range of applicability) was not reduced as compared with the Maximin-PP on its own.

Simplicity The combination of the Rights-Threshold Principle and Maximin-PP is less simple than the Maximin-PP on its own: we have at least the concept of rights-violations in addition, and also the concept of lexical priority. This raises the technical apparatus from seven to nine concepts.

Above, I argued that adopting a rights threshold provides a unifying rationale because most if not all relevant cases of harm to the environment and/or to human health will be subsumable under it—as will be cases of threat of catastrophe. This argument is interesting from an RE perspective: arguably, this means that the rights threshold has more **unifying power** than other alternatives. This is a theoretical virtue, but one that was not selected as relevant in the initial setup. Still, it distinguishes the rights threshold from the other candidates, and clearly seems to speak in its favor.

8.2.5 *Adopting the Rights-Maximin-PP*

At the end of Step A_5, I am now adopting the "Rights-Maximin Precautionary Principle for Combinations of Uncertainty and Incommensurability (Rights-Maximin-PP)" as the current system. It can better account for current commitments than available alternatives, and its theoretical virtuousness was not significantly decreased compared with the Maximin-PP, which was chosen as the current system at the last step. The Rights-Maximin-PP consists of the following two parts:

Maximin-PP for Combinations of Uncertainty and Incommensurability (Maximin-PP) Select the course of action with the best worst case if you are either:

- In a situation of decision-theoretic risk or uncertainty (or some combination), and the outcomes of the available actions can be ranked on an ordinal scale, and all courses of action alternative to the one selected by maximin have outcomes that are incommensurably worse than the best worst case; or
- In a situation of (partial) decision-theoretic uncertainty, outcomes can be ranked on a cardinal scale, and all courses of action alternative to the one selected by maximin have negative outcomes that outweigh every potential gain that could be made above the level that can be guaranteed by maximin.

The Rights-Threshold Principle Threats of rights violations have lexical priority over other threats, and are incommensurable with chances of other kinds of gains.

In the next step, current commitments are adjusted with respect to the Rights-Maximin-PP.

8.3 Step B_5: Adjusting Commitments to the Rights-Maximin-PP

Two commitments that are in tension with the Rights-Maximin-PP can be adjusted, and a conflicting commitment is given up (Sect. 8.3.1). But when searching for further relevant commitments, problems for the Rights-Maximin-PP emerge: relevant information about possible outcomes should not be irrelevant for the decision-process (Sect. 8.3.2).

8.3.1 Trying to Increase Account

The two commitments to giving priority to human health and the environment can be adjusted in order to increase their agreement with the current system. That is, from

EC 23 *Pro tanto*, threats of harm to human health have lexical priority for precaution. [low] [emerged at Step B_4]

to

C 7 Threats to human health have lexical priority for taking precautionary measures insofar as they are threats of rights violations. [high] [replaced EC 23 at Step B_5]

And I change the commitment:

EC 24 *Pro tanto*, threats to the environment have lexical priority for precaution. [low] [emerged at Step B_4]

to:

C 8 Threats to the environment have lexical priority for taking precautionary measures insofar as they are threats of rights violations. [high] [replaced EC 24 at Step B_5]

Arguably, by adjusting the commitment in this way, a lot of the original intention of the commitment is preserved, namely, that threats to the environment deserve special attention. At the same time, it makes sense to adjust the weight of this commitment from "low" to "high", since we can now better defend this commitment by being able to cite a *reason* for why some threats to the environment have lexical priority.

Then we have a commitment that is in direct conflict with the current system, by demanding that no threat is given priority insofar as it threatens a specific entity.

EC 1 The target PP is neither restricted to threats to specific entities (e.g., the environment and/or human health), nor is there a category of threat that takes lexical priority for the application of a PP insofar as it is a threat to specific entities. [low] [emerged at Step B_1]

I argue that this commitment can be rejected on the basis that the current system, the Rights-Maximin-PP (S_4), shows how, by accepting that if we take a category of threat to have lexical priority, we gain a lot in terms of account, applicability, and determinacy. Also, the weight of this commitment is only *low*—it was more a working hypotheses than a substantial commitment.

C 9 Non-EC 1 [replaced EC 1 at Step B_5]

8.3.2 Searching for Further Relevant Commitments

So far, I treated the Maximin-PP more or less as "set", i.e., as being in equilibrium with the relevant commitments that it is supposed to systematize. The focus was on how the value, or respectively evaluative, commitments can be systematized by adding a threshold of lexical priority. Starting at Step A_5, I compared candidates for a part of the system that can supplement the Maximin-PP in order to arrive at a target-system that meets the pragmatic-epistemic objective. But when moving on, we need again to take the whole system into perspective. In this subsection, I explore whether there are further relevant commitments that would destabilize the current position.

So far, I have bracketed the question of what counts as "reasonable outcomes", i.e., which outcomes are still plausible enough to include when considering alternative courses of action, thinking that this is a problem of risk assessment and not relevant for the choice of the decision-principle.[4] However, when consulting the literature on maximin principles and precaution, it emerges that this is something that needs to be taken seriously. Take the following example:

> [When] deciding how to arrange the ventilation in my house, I take into account that insects may try to enter through certain types of ventilators, but I disregard remote possibilities such as that a tropical snake from the nearby zoo tries to break in through the ventilator. The line has to be drawn somewhere, but there is no general rule telling us exactly where to draw it in different decision problems. (Hansson 2003, 296)

As Hansson (2003) argues, using a maximin approach transfers the difficulties from the analysis of a problem to the prior construction of a formal decision problem. Identifying what the *relevant* worst case is far from trivial (Betz 2010; Roser 2017, 1402). It is true that every decision principle for decisions under uncertainty faces the problem of how to identify reasonable outcomes, as Gardiner (2006) argues. But since maximin principles focus almost exclusively on worst cases, they are especially sensitive to how the decision problem is framed and where we draw the line. I thus adopt the following commitment:

EC 28 A decision principle for decisions under uncertainty needs criteria to decide which outcomes should still be included as "reasonable" or "plausible" enough. [medium] [emerged at Step B_5]

A further problem for maximin principles is stressed by Roser (2017, 1402):

> If our evidence is such as to allow for a judgement about the realistic range of consequences, this same evidence surely allows for at least some [comparisons of likelihood, T.R.] within and beyond that range.

While I do not want to follow Roser in his specific use of "epistemic probabilities", I do agree that if we have enough information to decide which outcomes to include as realistic enough, then this information should not simply be discarded when deciding which course of action we should choose. This is also in line with my commitment that the price of precaution should be proportional not only to the seriousness, but also to the plausibility of a threat:

EC 19 The price of precaution should be proportional to the seriousness and the plausibility of the threat, given the available alternatives. [low] [emerged at Step B_2]

---

[4] See Sect. 6.4.2, p. 134, where I stipulate that we have something in the background that allows us to distinguish plausible outcomes from those that are not plausible.

If the evidence in favor and against the possibility of an outcome does not play a role beyond deciding whether an outcome is "reasonable" or not, then it is hard to identify measures that really are proportional to the plausibility of a threat.[5] I thus endorse the following emerging commitment:

EC 29 The information we have about possible outcomes of courses of actions should not be irrelevant for the decision process only because it is not sufficient to assign reliable probabilities. [medium] [emerged at Step B_5]

8.3.3 *The Adjusted Set of Current Commitments, C_5*

As Fig. 8.2 shows, adjusting EC 23, EC 24, and EC 1 to C 7, C 8, and C 9 did increase the account value for the Rights-Maximin-PP. However, the two emerging commitments EC 28 and EC 29 decrease it again.

8.4 Step A_6: From Rights-Maximin-PP to Rights-TPA

The Tripartite Precautionary Approach (TPA) is adjusted to also account for my commitments to giving priority to human rights. The resulting "Tripartite Precautionary Approach to Threats of Rights Violations (Rights-TPA)" is then compared with the Rights-Maximin-PP both with respect to their ability to account for commitments (Sect. 8.4.1) and their theoretical virtues (Sect. 8.4.2). In particular, two emerging commitments from Step B_5 make the TPA more attractive than the Maximin-PP, since they do directly conflict with the latter (and not only with some of its implications). Consequently, the Rights-TPA is adopted at the end of Step A_6 Sect. 8.4.3).

In this step, I argue that the two emerging commitments from Step B_5 now make the TPA more attractive than the Maximin-PP: they are commitments that do not only conflict with some of the implications of the current system, S_5, but that conflict directly with one of its central parts, the Maximin-PP. As I will argue in the following, adapting the *Tripartite Precautionary Approach (TPA)* to the idea of threats of rights violations having lexical priority does avoid these problems and is,

[5] Unless we understand "plausible" as a yes/no question, like whether an outcome is "reasonable" or "realistic", and a measure is proportional if it is taken against a plausible threat and not proportional if the threat is not plausible. But this neither seems convincing nor is it how I introduced "plausibility" in Sect. 6.4.2.

Step B_5: Current Set of Commitments C_5

General Commitments on Precaution and Precautionary Decision-Making

	C1	C2	IC3	IC4	IC5	IC6	IC7	IC8	C9	EC2	C5	C6	C3	EC16	EC17	EC18	EC19	EC20	EC21	EC22
TPA	✓2	✓2	✓6	-1	✓6	✓6	-1	✓2		✓2	-0.5	✓2	✓4	-1.5	-1	✓2	✓2	✓4	✓4	✓4
Maximin-PP	-0.5	-0.5	-1.5	-1	✓6	✓6	-1	✓2		-0.5	✓2	-0.5	✓4	-1.5	-1	✓2	✓2	✓4	✓4	✓4
Rights-Max.-PP	-0.5	-0.5	-1.5	-1	✓6	✓6	✓4	✓2	✓2	-0.5	✓2	-0.5	✓4	-1.5	-1	✓2	✓2	✓4	✓4	✓4

	C7	C8	EC25	EC26	EC27	EC28	EC29
TPA			-0.5	-1.5	-0.5		
Maximin-PP			-0.5	-1.5	-0.5		
Rights-Max.-PP	✓2	✓2	✓2	✓6	✓2	-1	x -4

Commitments on Toy-Examples

	IC9	IC10	IC11	C4	IC13	IC14	IC15	IC16	IC17	IC18	IC19	IC20	IC21	IC22
TPA	✓6	✓6	✓4	-1	✓6	✓6	✓4	✓4	✓4	✓6	✓4	✓6	✓4	✓6
Maximin-PP	✓6	✓6	✓4	-1	✓6	✓6	✓4	✓4	✓4	✓6	✓4	✓6	✓4	✓6
Rights-Max.-PP	✓6	✓6	✓4	-1	✓6	✓6	✓4	✓4	✓4	✓6	✓4	✓6	✓4	✓6

Commitments concerning Precaution, Climate Change, and Solar Radiation Management (SRM)

	IC23	IC24	IC25	IC26	IC27	IC28	IC29	IC30	IC31
TPA	✓6	~2	✓4	✓4	✓4	✓6	✓4	✓4	✓6
Maximin-PP	✓6	~2	✓4	✓4	✓4	✓6	✓4	-1	✓6
Rights-Max.-PP	✓6	✓4	✓4	✓4	✓4	✓6	✓4	-1	✓6

✓ full account	+2
~ partial account	+1
consistent, but no account	-0.5
× Conflict	-2
black: high weight	* 3
grey: medium weight	* 2
white: low weight	* 1

Fig. 8.2 End of Step B_5: current commitments C_5

overall, a more convincing candidate system—i.e., it better fulfills the RE criteria with respect to the input commitments and the pragmatic-epistemic objective.

In part, I arrived at combining the *Rights-Threshold Principle* with the Maximin-PP because the latter, with its incommensurability criterion, seemed to lend itself to an interpretation along the lines of certain outcome values having lexical priority over others. However, the TPA is at least as well suited for such a combination: the "harm condition" of its Precautionary Tripod in the sense of a failure of meeting a "safety target" is well suited for the idea of lexical priority of rights. I propose the following adaptation of the TPA:

The **Tripartite Precautionary Approach to Threats of Rights Violations (Rights-TPA)**:

- The **Rights Meta Precautionary Principle (Rights-MPP)**: Uncertainty must not be a reason for inaction when there are threats of rights implications.
- The **Precautionary Tripod**: The elements that have to be specified in order to obtain an action-guiding precautionary principle version: If there is a threat that meets the *harm condition* (i.e., a specific rights violation) under a given *knowledge condition* then a *recommended precaution* should be taken.
- **Proportionality**: Demands that the elements of the Precautionary Tripod are adjusted proportionally to each other, understood as *Consistency*: The recommended precaution must not be recommended against by the same PP version, and *Efficiency*: Among those precautionary measures that can be consistently recommended by a PP version, the least costly one should be chosen.
  The **starting point for a rights-based PP version**: If there is (1) a threat of a wrongful rights violation, then (2) select the least stringent knowledge condition that results in a consistently applicable version of PP given the harm condition. To comply with the Rights-MPP, uncertainty must neither render the PP version inapplicable nor lead to continual delay in taking measures to prevent rights violations.

It is noteworthy that the Rights-TPA, in itself, does not tell us which rights implications are permissible and which are wrongful rights violations. The Rights-MPP only tells us that uncertainty must not lead to inaction when there are threats of rights implications. But this does not exclude that there are other reasons than uncertainty that make the rights implication acceptable or permissible—e.g., consent of the party on whom the threat is imposed might make a difference. To clarify this is, however, the subject of a theory of rights.[6] The Rights-TPA only tells us that if there is a threat of rights implications, and all reasons for inaction aside from uncertainty have been ruled out (i.e., if it were a wrongful rights violation,

[6] By this "move", I hope to avoid discussions about, e.g., why driving a car is permissible even though you impose a very low threat of dying on everyone. Why this is still permissible—e.g., because everyone, even those who do not drive themselves, benefit from the practice of car-driving—has to be explained by a theory of rights. (A more detailed analysis of the threat impositions in car driving might, however, also reveal that it cannot be prohibited by a proportional Rights-PP-Version.)

would it materialize), *then* the uncertainty must not lead to inaction: the Rights-TPA requires us to find a Rights-PP-Version that consistently can recommend action.

Comparing Rights-TPA and Rights-Maximin-PP In Step A_5, I treated the Maximin-PP as *set* and was searching for a plausible candidate that could supplement the Maximin-PP as a lexical priority threshold. I argued that giving lexical priority to avoiding threats of rights violations is one of the most plausible candidates. Now, we are keeping this part—the rights threshold—constant, and are comparing whether the Maximin-PP or the TPA is better suited to complement it as a decision-making approach.

8.4.1 Rights-Maximin-PP and Rights-TPA: Account

Contrary to the Rights-Maximin-PP, the Rights-TPA can account for EC 28 and EC 29:

EC 28 A decision principle for decisions under uncertainty needs criteria to decide which outcomes should still be included as "reasonable" or "plausible" enough. [medium] [emerged at Step B_5]

The (Rights-)TPA avoids the problem of "reasonable outcomes" by demanding that a precautionary measure against a threat should at least meet the same knowledge condition; and demanding that the least stringent knowledge condition should be chosen that still leads to a consistently applicable PP version. Understood in this way, "reasonable" outcomes are those against which we can reasonably take precautions. This answer to the "reasonable outcomes"-problem does thereby not consist in adding some *de minimis* condition, i.e., adding some more or less arbitrary threshold for how likely outcomes have to be in order to be included (Steel 2015, 37).

EC 29 The information we have about possible outcomes of courses of actions should not be irrelevant for the decision process only because it is not sufficient to assign reliable probabilities. [medium] [emerged at Step B_5]

Contrary to the Maximin-PP, where evidence only plays a role in determining which outcomes should be included, the Rights-TPA takes available evidence into account when deciding on a course of action: the TPA can operate both with quantitative knowledge conditions, e.g., numerical probabilities, and with qualitative rankings of knowledge conditions, i.e., ordinal rankings (Steel 2015, 6; 111). Examples of knowledge conditions that Steel mentions are, e.g., probability thresholds of 34%, 50%, or 10% (Steel 2015, 202), which are quantitative knowledge conditions expressed in numerical probabilities. However, other examples are "hypothetically possible" which is less stringent than "a scientifically established mechanism type exists that could bring the outcome about", which is again less stringent than there being "a known specific scientific mechanism observed to be in operation likely to lead to a specific outcome" (Steel 2015, 113). This means that as long as knowledge conditions of a harm condition and about the outcomes of a precautionary measure can at least be ordinally ranked, they can be taken into account and compared when deciding on a proportional precautionary response that is required by a threat (of rights violations).

For the rest of the commitments, I don't assess in detail for each of them whether or not the candidates can account for them—this would require a lot of work in terms of specifying a lot more background information about which rights might be at stake, etc. But it seems plausible enough that the Rights-TPA will be able to account for more commitments than the Rights-Maximin-PP, *if* the background information *were* specified accordingly.

Take the examples of the cases *Asbestos 1* and *Asbestos 2*:

Case 5: Asbestos 1 Large-scale mining and manufacturing of asbestos has started about 15 years ago. Asbestos is seen as a desirable material because of its properties like sound absorption, tensile strength, and its resistance to fire and heat. Production costs are low, so it is also affordable. However, there are observations and reports that associate lung diseases with inhaling asbestos, although no systematic scientific research has been done on it so far; thus, a clear connection cannot be proved, and the diseases might have other causes.

We have to choose between the following four options:

(i) BAU: Continuing business-as-usual,

(ii) Research: Starting systematic scientific research on the harmfulness of asbestos dust, including long-term studies and mortality statistics of asbestos workers,

(iii) Research&Regulation: Starting systematic scientific research while already strictly regulating asbestos production, including, e.g., limiting exposure of workers to asbestos dust, and making compensation arrangements, based on agreed liabilities, or

(iv) Ban: Banning asbestos.

Case 6: Asbestos 2 Large-scale mining and manufacturing of asbestos has started about 45 years ago. Asbestos is seen as a desirable material because of its properties like sound absorption, tensile strength, and its resistance to fire and heat. Production costs are low, so it is also affordable. It is widely used in a range of applications, and its use is continuing to grow. However, it is now accepted that the inhalation of asbestos dust can cause a lung disease called "asbestosis".[7] Recently there have been cases of asbestosis that have been complicated by lung cancer, but a clear connection is difficult to prove, one reason being that smoking has become increasingly popular

[7] E.g., a health study of asbestos workers has shown that 66% of those employed for 20 years or more suffered from asbestosis, versus none of those employed for less than 4 years (Harremoës et al. 2001, 54).

and is also seen as a potential cause for lung cancer.[8] Additionally, some concerns have been raised that the inhalation of asbestos dust might cause other long-latent-period harm to people. There are other, presumably safer substances available, but they are much more expensive in production costs.[9]

We have to choose between he following three options:

(i) BAU: Continuing business-as-usual,

(ii) Research&Regulation: Starting systematic scientific research while already strictly regulating asbestos production, including, e.g., limiting exposure of workers to asbestos dust, and making compensation arrangements, based on agreed liabilities,

(iii) Ban: Banning asbestos.

If we assume that the threshold of the "Minimally Acceptable Level", as specified in Figs. A.3, p. 262, and A.4, p. 263, refers to threats of wrongful rights violations (e.g., the right of the workers and consumers to human health), then the Rights-TPA can perfectly account for the two commitments in these cases. In case *Asbestos 1*, two courses of action have worst cases that do not meet the harm condition: (iii), Research&Regulation, and (iv), Banning Asbestos. So both these options can be consistently recommended by the Rights-TPA. However, option (iii) is the less costly option, so efficiency as part of the proportionality criterion of the Rights-TPA will tell us to choose option (iii). This fits with my commitment:

IC 15 In case 5, *Asbestos 1*, we should choose option (iii), Research&Regulation. [medium]

The Rights-Maximin-PP, however, cannot tell us whether we should choose (iii) or (iv) in *Asbestos 1*, because not "all courses of action alternative to the one selected by maximin have outcomes that are incommensurably worse than the best worst case". It thus cannot account for the commitment, even though it is consistent with it.

In case *Asbestos 2*, only one course of action has a worst case that does not meet the harm condition: (iii), Banning Asbestos. Consequently, the Maximin-PP tells us to choose it, and can thereby account for the commitment. But so does the Rights-TPA. Consequently, the TPA tells us to choose it, which is again in agreement with my commitment:

IC 16 In case 6, Asbestos 2, we should choose option (iii), banning asbestos and substituting it with other, safer substances. [medium]

---

[8] I omit here that in Germany, before smoking became popular and while lung cancer rates were still relatively low, a connection between asbestos and lung cancer was already accepted in 1938 (Harremoës et al. 2001, 54).

[9] For reasons of simplicity, I do not consider what kinds of measures were already taken, and how effective (or not) they have been.

That is, not only can the Rights-TPA account for the two commitments, it can also account for the difference between the two cases, i.e., why once Research&Regulation is chosen over Banning Asbestos, while in the other case Banning Asbestos should be chosen over the Research&Regulation option. Since the Maximin-PP has no such efficiency criterion, it fails to account for the commitment concerning *Asbestos 1*, and also for the difference between the two cases.

In Fig. 8.3, the results from assessing account for current commitments are summarized.[10] There are now some interesting trade-offs in terms of for which commitments each candidate can account: the TPA and the Maximin-PP both fail to account for C 7, C 8, and EC 25–EC 27, which all are moral value-commitments. Their rights-based adaptations, the Rights-Maximin-PP and the Rights-TPA, both can account for these commitments but have other problems: the Rights-Maximin-PP (like the Maximin-PP on its own) can't account for EC 28 and EC 29 which concern the role of evidence for a PP. The Rights-TPA can account for these commitments, but fails to account for commitments concerning individual risk-taking, e.g., IC 9–IC 11. All in all, the Rights-TPA still reaches the highest account value, namely 152.5, whereas the Rights-Maximin-PP reaches 144.

8.4.2 *Rights-Maximin-PP and Rights-TPA: Theoretical Virtues*

When assessing Determinacy and Practicability for the Maximin-PP and the TPA in Step A4, the Maximin-PP did rank higher than the TPA. Now both these candidates have been supplemented with a rights threshold, but since this threshold is the same for both candidates, it makes no difference for the comparative assessment of Determinacy and Practicability. Consequently, the Rights-Maximin-PP will rank higher than the Rights-TPA with respect to these virtues. This leaves us with assessing scope and simplicity.

Scope While the range of applicability of the Rights-TPA is the same as the one of the Rights-Maximin-PP, it has a broader application-set, i.e., there are more situations in which it will yield an action-guiding verdict. While this is not directly relevant for scope in the sense as I understand and use it here, it is relevant because this broader application set actually allows the Rights-TPA to *account* for more commitments. For example, the TPA does not focus on the best worst case, but on how to most efficiently avoid or reduce threats of not meeting a defined safety target (i.e., in the case of the Rights-TPA, this safety target is not violating (specific) rights). This means that it can sensibly be applied when several "worst cases" would meet the safety target: it then takes benefits and costs into account by demanding that the most efficient precautionary measure should be taken.

10 Please note that for case-specific commitments, it has been stipulated that there is a theory of rights in the background that yields outcome evaluations that fit with the commitments.

Step A_6: Account for Commitments C_5

General Commitments on Precaution and Precautionary Decision-Making

	C1	C2	IC3	IC4	IC5	IC6	IC7	IC8	C9	EC2	C5	C6	C3	EC16	EC17	EC18	EC19	EC20	EC21	EC22
TPA	✓2	✓2	✓6	-1	✓6	✓6	-1	✓2	-0.5	✓2	-0.5	✓2	✓4	-1.5	-1	✓2	✓2	✓4	✓4	✓4
Maximin-PP	-0.5	-0.5	-1.5	-1	✓6	✓6	-1	✓2	-0.5	-0.5	✓2	-0.5	✓4	-1.5	-1	✓2	✓2	✓4	✓4	✓4
Rights-Max.-PP	-0.5	-0.5	-1.5	-1	✓6	✓6	✓4	✓2	✓2	-0.5	✓2	-0.5	✓4	-1.5	-1	✓2	✓2	✓4	✓4	✓4
Rights-TPA	✓2	✓2	✓6	✓4	✓6	✓6	✓4	✓2	✓2	✓2	-0.5	✓2	✓4	-1.5	-1	✓2	✓2	✓4	✓4	✓4

	C7	C8	EC25	EC26	EC27	EC28	EC29
TPA	-0.5	-0.5	-0.5	-1.5	-0.5	✓4	✓4
Maximin-PP	-0.5	-0.5	-0.5	-1.5	-0.5	-1	× -4
Rights-Max.-PP	✓2	✓2	✓2	✓6	✓2	-1	× -4
Rights-TPA	✓2	✓2	✓2	✓6	✓2	✓4	✓4

✓ full account	+2
~ partial account	+1
consistent, but no account	-0.5
× Conflict	-2
black: high weight	* 3
grey: medium weight	* 2
white: low weight	* 1

Commitments on Toy-Examples

	IC9	IC10	IC11	C4	IC13	IC14	IC15	IC16	IC17	IC18	IC19	IC20	IC21	IC22
TPA	✓6	✓6	✓4	-1	✓6	✓6	✓4	✓4	✓4	✓6	✓4	✓6	✓4	✓6
Maximin-PP	✓6	✓6	✓4	-1	✓6	✓6	✓4	✓4	✓4	✓6	✓4	✓6	✓4	✓6
Rights-Max.-PP	✓6	✓6	✓4	-1	✓6	✓6	✓4	✓4	✓4	✓6	✓4	✓6	✓4	✓6
Rights-TPA	-1.5	-1.5	-1	-1	✓6	✓6	✓4	✓4	✓4	✓6	✓4	✓6	✓4	✓6

Commitments concerning Precaution, Climate Change, and Solar Radiation Management (SRM)

	IC23	IC24	IC25	IC26	IC27	IC28	IC29	IC30	IC31
TPA	✓6	~2	✓4	✓4	✓4	✓6	✓4	✓4	✓6
Maximin-PP	✓6	~2	✓4	✓4	✓4	✓6	✓4	-1	✓6
Rights-Max.-PP	✓6	✓4	✓4	✓4	✓4	✓6	✓4	-1	✓6
Rights-TPA	✓6	✓4	✓4	✓4	✓4	✓6	✓4	✓4	✓6

Account-Values:

Maximin-PP: 119
TPA: 152
Rights-Maximin-PP: 145.5
Rights-TPA: 144

Fig. 8.3 Step A_6: account for commitments C_5

And if a precaution against one sort of rights violation threatens another kind of rights violation, the Rights-Meta-PP will again demand action—i.e., every threat gets addressed through an iterative application of PP-versions. This allows for rights being hierarchical, too, and, e.g., giving more priority to avoiding threats to very fundamental rights without making threats of violations of more "minor" rights irrelevant. An example would be a case where a threat to a fundamental right is addressed by a precautionary measure that threatens a more minor right. This latter threat that is caused by the precautionary measure does not itself meet the harm- and knowledge condition combination of the PP-version used to justify the precautionary measure, so it does not cause a problem for consistency. However, the Rights-Meta-PP demands that also with respect to this other threat of a rights violation, uncertainty must not lead to inaction.[11]

Simplicity The TPA does not need additional criteria for reasonable outcomes because this is addressed as a part of proportionality—i.e., it emerges organically from the candidate system. Even though the current version of the Maximin-PP does not have such a reasonable outcomes criterion and we therefore cannot assess how simple it would be, from the structure of the Maximin-PP it is hard to imagine that such a criterion could be similarly integrated. I thus rank the Rights-TPA as simpler than the Rights-Maximin-PP.

8.4.3 Overall Comparison: Rights-Maximin-PP vs. Rights-TPA

The Rights-TPA can account for more commitments than the Maximin-PP. It also does not require an additional criterion for "reasonable outcomes", since this is built into the proportionality criterion of the TPA: "reasonable outcomes" are those against which we still can take precautionary measures that do not themselves meet the harm and knowledge condition of the threat. This makes an additional criterion superfluous.

On this basis, I adopt the Rights-TPA as the new current system.

[11] And if taking an action that involves some more minor threat is the only way to address a more severe threat, then it seems plausible to argue that the reason for inaction with respect to the minor threat is not uncertainty.

8.5 Step B_6: Adjusting Commitments to the Rights-TPA

Commitments are adjusted to increase their agreement with the Rights-TPA (Sect. 8.5.1). First, a commitment that was already adjusted in Step B_3 is again adjusted in a different way. Second, the commitments concerning individual risk-taking are not accounted for by the Rights-TPA. It is argued that these commitments can defensibly be adjusted to be in agreement with the Rights-TPA: to meet the objective of formulating a defensible moral precautionary principle, it is more important to give a satisfying answer to what precaution requires in other-regarding contexts, than to formulate a more unifying approach that covers both classes of situations. The resulting set of commitments is summarized in Sect. 8.5.2.

8.5.1 Trying to Increase Account

One of the commitments the current system S_6, the Rights-TPA, cannot account for (but is consistent with) is the following:

C 4 The claim that the measles/mumps/rubella (MMR) vaccine might cause autism does not constitute a plausible threat. [medium] [replaced IC 12 at Step B_3]

This commitment is an adjustment of the following input-commitment:

IC 12 The claim that the measles/mumps/rubella (MMR) vaccine might cause autism is not a reason not to vaccinate your child. [medium]

I argued that it is unclear exactly what IC 12 expresses, and proposed to interpret it in the manner of C 4. However, in light of the current position, given the Rights-TPA, another interpretation is more convincing:

C 13 Not vaccinating your child is not a proportional precautionary measure against the alleged threat that the measles/mumps/rubella (MMR) vaccine might cause autism. [medium] [replaced C 4 as a replacement for IC 12 at Step B_6]

This is still closely connected to the claim expressed in IC 12, but at the same time in agreement with the Rights-TPA.

Adjusting the Subject Matter: Excluding Individual Risk-Taking Next, we have a whole subset of commitments that are not accounted for by the Rights-TPA. These are commitments concerning individual risk taking, like the "Job-Offers" example (see also the discussion in Sect. 7.5):

Case 9, Job Offers Suppose you live in New York City and are offered two jobs at the same time. One is a tedious and badly paid job in New York City itself, while the other is a very interesting and well-paid job in Chicago. But the catch is that, if you wanted the Chicago job, you would have to take the plane from New York City to Chicago (e.g., because this job would have to be taken up the very next day). Therefore there would be a very small but positive probability that you might get killed in a plane accident (example from Harsanyi 1975, 595).

IC 11 In case 9, *Job Offers*, you should choose the job in Chicago. [medium]

However, it makes sense that the commitment concerning Job Offers is actually a weaker one than the one expressed by IC 23, i.e., the commitment that is relevant for the subject matter should rather be:

C 10 In Case 9, *Job Offers*, the target system should not tell you to choose the job in New York. [high] [replaced IC 11 at Step B_6]

C 10 is implied by IC 11, though much weaker. But it is enough to capture the main function that IC 23 was intended to have: to make sure that the target system does not lead to clearly irrational decisions (where I am committed to that, all else being equal, choosing the job in New York would be irrational).

I argue that similar commitments about individual risk-taking can be adjusted in the same way: they are now, as a result of this RE process, no longer a part of the application set of the current system, and thereby are excluded from the relevant subject matter. This is not to say that precaution is not required or not possible when taking individual decisions that affect only oneself. But it expresses that precaution requires something *different*, something *more*, when making decisions that will (potentially) affect others and not just oneself. And these other-regarding contexts were the specific focus of this pragmatic-epistemic project. To meet my objective, it is thus more important to give a satisfying answer to what precaution requires in other-regarding contexts, than to formulate a more unifying approach that covers both classes of situations and unifies them under one systematic approach.

Thus, the other current commitments concerning individual precaution can be replaced analogously to IC 11:

C 11 The target system should not tell you not to wear protective clothing when making soap. [high] [replaced IC 9 at Step B_6]

C 12 In Case 11, *Worst Case Being Shot*, the target system should not tell you to choose option A. [high] [replaced IC 10 at Step B_6]

8.5.2 *The Adjusted Set of Current Commitments, C_6*

The current commitments at the end of Step B_6 are summarized in Fig. 8.4. By re-adjusting IC 12 from C 4 to C 13, and by excluding situations concerning individual risk-taking from the subject matter, the account value of the Rights-TPA could be increased from 152.5 to 156.5.

Step B_6: Current Set of Commitments C_6

General Commitments on Precaution and Precautionary Decision-Making

	C1	C2	IC3	IC4	IC5	IC6	IC7	IC8	C9	EC2	C5	C6	C3	EC16	EC17	EC18	EC19	EC20	EC21	EC22
TPA	✓2	✓2	✓6	-1	✓6	✓6	-1	✓2	-0.5	✓2	-0.5	✓2	✓4	-1.5	-1	✓2	✓2	✓4	✓4	✓4
Maximin-PP	-0.5	-0.5	-1.5	-1	✓6	✓6	-1	✓2	-0.5	-0.5	✓2	-0.5	✓4	-1.5	-1	✓2	✓2	✓4	✓4	✓4
Rights-Max.-PP	-0.5	-0.5	-1.5	-1	✓6	✓6	✓4	✓2	✓2	-0.5	✓2	-0.5	✓4	-1.5	-1	✓2	✓2	✓4	✓4	✓4
Rights-TPA	✓2	✓2	✓6	✓4	✓6	✓6	✓4	✓2	✓2	✓2	-0.5	✓2	✓4	-1.5	-1	✓2	✓2	✓4	✓4	✓4

	C7	C8	EC25	EC26	EC27	EC28	EC29
TPA	-0.5	-0.5	-0.5	-1.5	-0.5	✓4	✓4
Maximin-PP	-0.5	-0.5	-0.5	-1.5	-0.5	-1	× -4
Rights-Max.-PP	✓2	✓2	✓2	✓6	✓2	-1	× -4
Rights-TPA	✓2	✓2	✓2	✓6	✓2	✓4	✓4

Commitments on Toy-Examples

	C11	C12	C10	C13	IC13	IC14	IC15	IC16	IC17	IC18	IC19	IC20	IC21	IC22
TPA					✓6	✓6	✓4	✓4	✓4	✓6	✓4	✓6	✓4	✓6
Maximin-PP					✓6	✓6	✓4	✓4	✓4	✓6	✓4	✓6	✓4	✓6
Rights-Max.-PP					✓6	✓6	✓4	✓4	✓4	✓6	✓4	✓6	✓4	✓6
Rights-TPA	✓6	✓6	✓6	✓4	✓6	✓6	✓4	✓4	✓4	✓6	✓4	✓6	✓4	✓6

Commitments concerning Precaution, Climate Change, and Solar Radiation Management (SRM)

	IC23	IC24	IC 25	IC26	IC27	IC28	IC29	IC30	IC31
TPA	✓6	~2	✓4	✓4	✓4	✓6	✓4	✓4	✓6
Maximin-PP	✓6	~2	✓4	✓4	✓4	✓6	✓4	-1	✓6
Rights-Max.-PP	✓6	✓4	✓4	✓4	✓4	✓6	✓4	-1	✓6
Rights-TPA	✓6	✓4	✓4	✓4	✓4	✓6	✓4	✓4	✓6

✓ full account	+2
~ partial account	+1
consistent, but no account	-0.5
× Conflict	-2
black: high weight	* 3
grey: medium weight	* 2
white: low weight	* 1

Fig. 8.4 Step B_6: current set of commitments C_6

8.6 Step A_7 and B_7: Reaching Equilibrium?

In Steps A_7 and B_7, no adjustments to the position are made: the Rights-TPA remains the most convincing candidate from the available alternatives, and it is in agreement with the current commitments. This brings the reflective equilibrium process to a (preliminary) end point, and I analyze whether we have reached a justified position that is in a state of reflective equilibrium. I argue that contingent on the stipulations and simplifications made for the sake of the case study, the RE criteria are approximated.

Given the adjusted set of commitments, the Rights-TPA might not reach the highest ranking with respect to the RE criteria that would be *hypothetically* possible, but it reaches a high ranking, and since there is no plausible competitor available anymore, I argue that it is the best candidate for the target system. I am thus not making any adjustments to the system in Step A_7.

This leads us to another step of adjusting commitments, Step B_7. But the set of current commitments, C_6, was already ideally adjusted with respect to the Rights-TPA, which was S_6 and is also the current system, S_7. Thus, there are no further adjustments to be made that would increase the agreement between commitments and system.

In Chap. 3, I suggested that the RE process comes to an end when neither of the two steps brings any improvements with respect to the RE criteria. This is the **stopping rule**. It then has to be assessed whether a full RE state was reached by asking the following questions:

- Are the resulting commitments and the system in agreement?
- Can the position be supported by background theories?
- Does the system do justice to theoretical virtues?
- When comparing input commitments and resulting commitments, is it plausible that we have not abandoned the subject?
- Do (at least some of) the resulting commitments have independent credibility?
- Is the resulting position at least as plausible as available alternatives?

In the following, I discuss the answers to each of these questions in turn.

Agreement between Resulting Commitments and Resulting System? There are no conflicts between the resulting commitments and the resulting system. However, there are some commitments that are not accounted for, namely EC 17 and EC 16.

EC 17 The price of a precautionary measure consists of—compared with the course of action entailing the threat it is supposed to address—foregone benefits,[12] foregone opportunities, and additional threats. [medium] [emerged at Step B_2]

The (Rights-)TPA demands that in cases where more than one course of action can be consistently recommended, the least costly one should be chosen. However, what the "cost", or price, of a course of action is remains unspecified, and thus, while not conflicting with EC 17, the (Rights-)TPA cannot account for it. A possible way to change this would be to add the content of EC 17 to the system.

EC 16 The costs and responsibilities for precautionary measures should be distributed in a morally sound way. [high] [emerged at Step B_2]

While the Rights-TPA will prohibit some ways of distributing costs and responsibilities (if they threaten to violate rights), it does not provide a more general framework for the distribution of costs and responsibilities. With this commitment, it is possible to argue that it does not really belong to the subject matter of precaution and precautionary decision-making (as already hinted at in Sect. 6.6), but should rather be systematized by a theory of distributive justice or something similar. Most likely, the theory of rights that we already have to stipulate in the background for the Rights-TPA will have implications for these distributive questions, too.

Very importantly, the agreement between resulting commitments and resulting system is conditional on certain stipulations and simplifications that were made for the sake of the case study. The most important stipulation is the one that there is a sufficiently fleshed-out theory of rights in the background, that allows us to evaluate possible outcomes etc. in a way that fits with the evaluations in my commitments, and that allows the Rights-TPA to account for them.

Additionally, it would be necessary to search more systematically for potentially conflicting commitments, since only a small subset could be explicitly considered.

Thus, even if this criterion is fulfilled given the context of the case study, I am cautious not to assert that it is fulfilled all things considered.

Is the Position Supported by Background Theories? Whether or not there are conflicts with background theories, or respectively whether the resulting position can be supported by them, is something that still would have to be explored in depth. I am not doing this as part of the case study and thus can only point towards questions one could ask in order to assess it, e.g.: does the way that "rights" are used in the Rights-TPA fit with how it is used in other (moral) theories? Can threats be assessed in the way the Rights-TPA demands, i.e., does this fit with theories of risk assessment, epistemic theories about possibilistic knowledge, and similar? Even though I cannot address these questions here, initial work done on the connection between precaution and human rights suggests there is a good chance that these

[12] I take it that "foregone benefits" also includes direct monetary costs of precautionary measures that are spent, e.g., on installing safety measures, since the money used there cannot be spent for other purposes.

questions could be answered in a positive way in future work (see in particular Caney 2009; Roser 2020).

Does the System have Theoretical Virtues? Theoretical virtues have been extensively assessed during the case study, and we can conclude that the Rights-TPA does justice to the theoretical virtues that were selected as relevant in Chap. 5. This does not exclude the possibility that its virtuousness could be improved, or that further theoretical virtues may relevant. But, currently, it seems to fulfill the criterion to a satisfying degree, given the pragmatic-epistemic objective.

Input Commitments Respected/Subject Not Abandoned? Input commitments (initial and emerging) are IC 1–IC 31 and EC 1–EC 29. Resulting commitments are C 1–C 3, C 5–C 13, IC 4–IC 8, IC 13–IC 31, EC 2, EC 16–EC 22, EC 25–EC 29. Differences between the two are that in the resulting commitments

1. input commitments concerning what does or does not count as a precautionary measure—EC 5–EC 14—have been moved to the background as being explicated by *ExplicPrec*.
2. input commitments to specific actions in cases concerning individual risk-taking have been excluded, i.e., IC 9–11 have been replaced by C 10, C 11, and C 12.
3. some vague input commitments have been re-interpreted, i.e., from IC 1, IC 2, and IC12 to C 1, C 2, and C 13.
4. several input commitments have been adjusted with respect to the current system (at that time), namely EC 15, EC 3, EC 4, EC 23, EC 24, and EC 1 to C 3 and C 5–C 9.

When comparing the input and the resulting commitments, is it plausible that the subject matter was not abandoned, and that we did end up with a systematization of what we did set out to systematize?

I argue that, yes, this is plausible: each adjustment is defensible in light of the independent credibility of the adjusted commitment, the resulting position, and the pragmatic-epistemic objective.

(1) It is reasonable that what does or does not count as a precautionary measure is in the background to, but not part of, a position that concerns morally warranted precautionary actions and decisions. The resulting system, the Rights-TPA, does recommend measures that meet the criteria of being a precautionary measure, while not requiring the explication itself to be applicable.
(2) Excluding individual risk-taking, i.e., situations where only the agent themself is affected by the threats they impose on themself, can be defended with the argument that we are concerned with the question of what (other-regarding) morality demands of us in terms of precaution. Thus, there is an argument

referring to a plausible difference between those cases and other situations, that can be used to defend excluding them.[13]

(3) IC 1 and IC 2, the Rio and the Wingspread formulation of a PP, were already adopted as commitments with a low initial credibility, because I was aware that they are both vague and often contested. Thus, it was partly an expectation of the RE process that it would help to find an interpretation of these commitments that does them justice while being more plausible. Arguably, C 1 and C 2 fulfill these goals. The clarification from IC 12 to C 13 has a similar motivation and can be seen as providing an interpretation of the claim that a threat (autism from a vaccine, IC 12) is "not a reason" to avoid taking an action (vaccinating): namely, that not taking the action cannot be defended as a proportional precautionary measure (C 13).

(4) As for the other adjustments, only one of them consisted in a direct rejection of an input commitment (from EC 1 to C 9). The rest of them consisted in slight adjustments in order to increase account with the system, e.g., changing "irreversible" to "incommensurable" harm when replacing EC 3 by C 5, spelling out in a bit more detail what it means for a threat to be *plausible* when replacing EC 4 through C 6, or clarifying that threats to human health or the environment have lexical priority for precaution *insofar* as they are threats of rights violations when adjusting EC 23 and EC 24 to C 7 and C 8. None of these adjustments seems in danger of leading to a change of subject.

Lastly, that a substantial number of input commitments remained unchanged also lends support to the claim that the subject matter is still, in the relevant sense, "the same".

Independent Credibility of Resulting Commitments Independent credibility was not assessed in detail: from the start, I only assigned rough weights of *low–medium–high* to the commitments, loosely based on my reasons for adopting them. A substantial number of the input commitments "survived" the process—and since all the credibility that input commitments have is by definition independent of the RE process (because we hold them before the process starts), at least those resulting commitments that are also input commitments will have independent credibility.

At Least as Plausible as Available Alternatives? As part of the case study, alternatives were not developed and assessed in detail. To really defend the resulting position, it would be necessary to test it in further cases in order to explore whether we are willing to commit to its implications, and also to develop real alternatives, e.g., including another moral normative basis than rights, and to compare in detail which of them fulfills the RE criteria to a higher degree. But this is outside the scope of the current project, which in the first place is a case study for the application of reflective equilibrium. Such a study would also be beyond the powers of any single

[13] If we adopt the TPA as a broader approach to precautionary decision-making, and see the Rights-TPA as the relevant specification for other-regarding morality, we would also cover the individual risk-taking cases. This is not implausible.

epistemic agent to achieve, and thereby suggests that philosophy, and other cognitive practices, ultimately has to be seen as a collective project.

Contingent on the stipulations and simplifications made for the sake of the case study, I argue that the RE criteria are approximated at this point, and that a preliminary reflective equilibrium is reached.

8.7 Recapitulation Phase 3

The results of the steps of phase 3 are summarized in Fig. 8.5. I started by comparing different candidates for a normative threshold of lexical priority, and selecting the rights threshold to supplement the Maximin-PP (Step A_5). When commitments were adjusted with respect to the Rights-Maximin-PP (Step B_5), two emerging commitments destabilized the Rights-Maximin-PP as the current system, which led to the adoption of the Rights-TPA at Step A_6. After commitments were adjusted with respect to the Rights-TPA at Step B_6, the latter was again selected as the current system at Step A_7. Since adjusting commitments at Step B_7 did not result in any changes, the question was asked whether a reflective equilibrium was reached.

In the remainder of this section, I discuss the results of phase 3 with respect to reflective equilibrium in Sect. 8.7.1, and with respect to precautionary principles in Sect. 8.7.2.

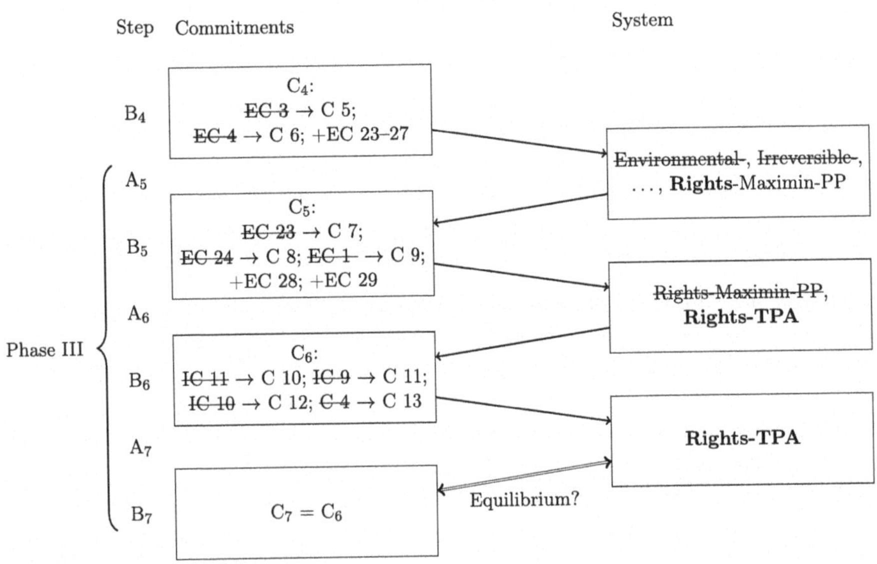

Fig. 8.5 Schematic overview of the steps of Phase 3

8.7.1 Results for Reflective Equilibrium

Main results from phase 3 for reflective equilibrium are:

- Working with stipulations and placeholders is sometimes unavoidable;
- Emerging commitments can play a decisive role, which is not a problem for the RE process;
- Fleshing out the position and making further relevant considerations explicit can be an important result of an RE process;
- It can be necessary to re-adjust already adjusted commitments;
- Whole subsets of commitments can be assessed and potentially excluded from the subject matter.

In phase 3, a preliminary RE state was reached. This equilibrium is contingent on certain stipulations in the background. While some of these stipulations are quite substantial—stipulating that there is a criterion for "reasonable outcomes" as was done in phases 1 and 2, or stipulating that there exists a suitable theory of rights—it does not seem unusual that at least *some* such stipulations and assumptions have to be made in an RE process: we have to start somewhere, which means that **sometimes we will just have to work with stipulations and place-holders** in the background in order to work out one position. Afterwards, of course, we should move on to address these stipulations—and depending on the outcome, this might again destabilize the position we reached. In the context of this RE implementation, one sensible way to continue would be to spell out the relevant sense of "uncertainty", since the assessment of theoretical virtues of the (Rights-)TPA has shown that the lack of a clear concept of "uncertainty" impairs the determinacy of the (Rights-) Meta-PP.[14]

In any case, assessing the resulting position with the RE criteria forces us to put the cards on the table, to admit weaknesses and unresolved issues, but also allows us to argue for why we see this position as defensible (see the analysis in Sect. 8.6). This is a positive result in favor of RE as a method of justification.

As further results from phase 3, **emerging commitments did play a decisive role**: that emerging commitments destabilize the Maximin-PP and support the Rights-TPA shows how RE is relative to those commitments that are explicitly considered. The process would most likely have taken a different path if those commitments had been made explicit from the beginning. The question is whether

[14] For example, (Steel 2015, chapter 5) develops a specific conception of "scientific uncertainty" to supplement his PP proposal.

this is a problem. I argue that it is not, because, firstly, the resulting set of commitments as a whole has to respect input commitments as a whole. So this is something that always has to be assessed with respect to the input commitments that are explicit at a current point in the RE process. This might mean that an adjustment that before could be reasonably seen as respecting input commitments is no longer defensible given further emerging commitments. But for the resulting commitments, it does not matter at which point an input commitment entered the process: they have to be respected in a way that makes it plausible that the subject was not abandoned, and that their independent credibility was not unwarrantably discarded. Thus, maybe we will take some "loops" that are in some sense "unnecessary", because adjustments that were made with respect to a subset of the input commitments later turn out not be defensible. But at the same time, such "loops" might be necessary to uncover further relevant commitments. **Fleshing out our set of commitments**, and becoming aware about further relevant considerations, **can also be an important result of an RE process**. It just also means that at an RE endpoint it is especially relevant to consider whether all relevant input commitments have been made explicit and are respected—and that there is always the possibility that further emerging commitments might destabilize our position. But this is in line with the general notion of justification via RE being preliminary.

That **respect for input commitments can depend on how the position develops** is demonstrated by the re-adjustment of IC 12, which at Step B_3 was replaced by C 4, but at Step B_6 this replacement got re-assessed and C 4 as a replacement for IC 12 was replaced by C 13.

IC 12 The claim that the measles/mumps/rubella (MMR) vaccine might cause autism is not a reason not to vaccinate your child. [medium]

C 4 The claim that the measles/mumps/rubella (MMR) vaccine might cause autism does not constitute a plausible threat. [medium] [replaced IC 12 at Step B_3]

C 13 Not vaccinating your child is not a proportional precautionary measure against the alleged threat that the measles/mumps/rubella (MMR) vaccine might cause autism. [medium] [replaced C 4 as a replacement for IC 12 at Step B_6]

This shows how the adjustment of an input commitment can be re-adjusted in light of the current position: the important point is that we are not simply going on to adjust C 4, but rather we go back to IC 12 and search for a better interpretation of this commitment in light of the current position.

Similar to phase 2, the **exclusion of a subset of commitments from the subject matter** was discussed at Step B_6. Contrary to commitments concerning cases where probabilities are available, the class of commitments concerning individual risk-taking ended up being excluded from the subject matter. However, this is defensible with respect to the pragmatic-epistemic objective, which is to formulate a moral precautionary principle, i.e., a principle for *other-regarding* decisions.[15]

8.7.2 Results for Precautionary Principles

Main results from phase 3 for precautionary principles are:

- A rights-based precautionary principle supplies a substantial justification for precautionary action, which is independent of whether or not there is a history of failed precaution;
- A rights-based precautionary principle can explain why some, but not all, threats to the environment or human health warrant lexical priority;
- However, the Tripartite Precautionary Approach (TPA) in its broader form can be acceptable independently of a specific moral theory, which might make it more suitable as a principle for public policy.

In the input commitments, I started out being committed to the claim that no class of threat takes lexical priority insofar as it is a threat to a specific entity. I committed to this because giving lexical priority to, e.g., harms to the environment seemed unduly narrow, and could lead to unacceptable trade-offs, e.g., accepting huge economic loss to avoid even insignificant damage to the environment (cf. Gardiner 2006, 45; Steel 2015, 84).

EC 1 The target PP is neither restricted to threats to specific entities (e.g., the environment and/or human health), nor is there a category of threat that takes lexical priority for the application of a PP insofar as it is a threat to specific entities. [low] [emerged at Step B_1]

By adopting the Maximin-PP, I accepted that there can be outcomes values that are incommensurable with other outcomes. This does not yet constitute a conflict with EC 1, since it leaves open whether harms that are incommensurable all concern harms to a specific entity. However, when continuing the process, giving lexical priority to threats of rights violations turned out to be a successful candidate for systematizing commitments—so successful that it made it defensible to reject EC 1. One can also debate whether giving lexical priority to avoiding rights violations

[15] One could also support this exclusion by adopting the Tripartite Precautionary Approach (TPA) as a broader approach to precautionary decision-making, which also covers individual risk-taking, and the Rights-TPA as a specific variant of the TPA for substantial moral decisions.

actually constitutes a conflict with EC 1, i.e., whether threats to rights are threats to a specific "entity" in the same way as threats to the environment or human health.

That the Tripartite Precautionary Approach to Threats of Rights Violations (Rights-TPA) turned out to be the most convincing candidate also supports the original approach of Steel (2015), which is shown to be a very comprehensive and systematic formulation of a precautionary principle. Making it a *Rights*-TPA is a possible way to make it more substantial as a moral principle, but we also have to acknowledge that this was not Steel's pragmatic-epistemic objective when formulating the TPA: throughout his book, Steel seems to conceive of the TPA as a principle for (regulatory) policy making, especially concerning the environment and human health (cf. Steel 2015, xi–xii, or the examples discussed on pp. 71–73). This makes it also comprehensible why Steel thinks that the harm condition is not something that can be determined by the TPA itself, but depends on value judgments:

> Decisions about the desired level of safety ultimately depend on value judgments that, ideally, would be generated from a deliberative democratic process that is sensitive to concerns of those who would be impacted by the decision. (Steel 2015, 201)

If the objective is to formulate and defend a principle for policy-making, then avoiding commitment to substantial moral values or theories is advisable because it makes the resulting system more broadly acceptable.[16]

Thus, the TPA might be more suitable as a basis for policy-making than a principle that is based on substantial moral commitments. However, it also leaves it open to a significant degree how the harm condition is set. By basing his Meta-Precautionary Principle on an historical argument referring to an historical pattern of significant errors in regulatory decisions at the expense of the environment and human health (Steel 2015, chapter 4), Steel avoids commitment to a specific ethical theory and achieves a principle that might be broadly acceptable. But this argument fails to explain *why* we should take precautionary action to protect the environment and human health *even if* no such history should exist. By adapting the Meta-PP to the Rights-Meta-PP, the justification of the resulting position is made independent of the historical argument. This does not mean that learning from history becomes irrelevant: the historical argument can still be relevant as background information when, e.g., threats are assessed—for example, because we have learned that threats to the environment often have long latent periods and might lead to almost irreversible system changes.

Compared with the Meta-PP, the *Rights*-Meta-PP is based on a substantial moral claim:

The Meta Precautionary Principle (MPP) Uncertainty must not be a reason for inaction in the face of serious (environmental) threats.

[16] Cf. Steel (2015, 93): "The argument for PP I develop here, then, has the attraction of avoiding reliance on debatable assumptions about ethical theory."

The Rights Meta Precautionary Principle (Rights-MPP) Uncertainty must not be a reason for inaction when there are threats of rights implications.

The reference to threats of rights implications serves at the same time as a powerful unifying rationale to explain why some, but not all, threats to the environment or human health warrant lexical priority. It can explain why rational choice theory with its indifference about risk-taking and risk-imposing situations often fails to capture the important normative basis for taking precautions (cf. Roser 2020).

It is noteworthy that the TPA and the Rights-TPA can to some degree coexist: unless the harm-condition of a PP-version of the TPA is set in a way that conflicts with the Rights-Meta-PP, they will not lead to conflicting verdicts. And if the harm condition of the TPA is defined in the way that Steel imagines—in a deliberative democratic process that is sensitive to the concerns of those affected by the decision—then it seems likely that it will typically be "triggered" at least by threats of grave rights violations.

But they both differ in the sense that the TPA can be acceptable independently of a specific moral theory or respectively that it is possible to supplement it with different moral theories like consequentialism or other deontological theories—whereas the Rights-TPA expresses (given an adequate theory of rights) determinative value judgments about which measures should be taken given which threats. Both can be defensible, depending on the input commitments and the pragmatic-epistemic objective of the epistemic agent.

Next, Chap. 9 discusses what we can learn from the case study for applying reflective equilibrium as a method.

References

Aldred J (2013) Justifying precautionary policies: incommensurability and uncertainty. Ecol Econ 96:132–140. https://doi.org/10.1016/j.ecolecon.2013.10.006

Arrow KJ, Fisher AC (1974) Environmental preservation, uncertainty, and irreversibility. Q J Econ 88(2):312–319. https://doi.org/10.2307/1883074

Betz G (2010) What's the worst case? The methodology of possibilistic prediction. Analyse und Kritik 32(1):87–106

Caney S (2009) Climate change and the future: discounting for time, wealth, and risk. J Soc Philos 40(2):163–186

Chang R (2013) Incommensurability (and incomparability). In: Lafollette H (ed) International encyclopedia of ethics. https://doi.org/10.1002/9781444367072.wbiee030

Gardiner SM (2006) A core precautionary principle. J Polit Philos 14(1):33–60

Hansson SO (2003) Ethical criteria of risk acceptance. Erkenntnis 59(3):291–309

Harremoës P, Gee D, MacGarvin M, Stirling A, Keys J, Wynne B, Vaz SG (eds) (2001) Late lessons from early warnings: the precautionary principle 1896–2000. Office for Official Publications of the European Communities, Luxembourg

Harsanyi JC (1975) Can the maximin principle serve as a basis for morality? A Critique of John Rawls's Theory. Am Polit Sci Rev 69(2):594–606. https://doi.org/10.2307/1959090

Hartzell-Nichols L (2012) Precaution and solar radiation management. Ethics, Policy and Environment 15(2):158–171. https://doi.org/10.1080/21550085.2012.685561

Hartzell-Nichols L (2013) From 'The' Precautionary Principle to Precautionary Principles. Ethics, Policy and Environment 16(3):308–320
Hartzell-Nichols L (2017) A climate of risk: precautionary principles, catastrophes, and climate change. Routledge, New York
Roser D (2009) A Baker's Dozen for future generations. University of Zurich, Zurich
Roser D (2017) The Irrelevance of the Risk-Uncertainty Distinction. Sci. Eng. Ethics 23(5):1387–1407. https://doi.org/10.1007/s11948-017-9919-x
Roser D (2020) Don't Look too far: rights as a rationale for the Precautionary principle. In: Akande D, Kuosmanen J, McDermott H, Roser D (eds) Human rights in the 21st century. Oxford University Press, Oxford, pp 305–322
Steel D (2015) Philosophy and the precautionary principle. Cambridge University Press, Cambridge
Sunstein CR (2007) The catastrophic harm precautionary principle. Issues in Legal Scholarship 6(3):1–29

Chapter 9
Results and Discussion: Justifying a Precautionary Principle as a Case Study for Reflective Equilibrium

In this book I have conducted an explicit and comprehensive case study of how one can apply reflective equilibrium (RE) as a method. Now it is time to take stock: what can we learn for RE and for how it can be used in philosophy? And what can we learn from applying RE for the subject of the case study, precautionary principles? In other words, this chapter addresses desideratum (6), which was identified in Chap. 1:

Desideratum 6 The application and its results should be evaluated and critically discussed in order to learn from it for the use of RE as a method.

As we saw in the introduction, there are two fundamental worries about RE as a method. On the one hand, RE is seen as *vacuous*, or too permissive. According to Foley, RE is at best meta-advice: because it neither tells you what data are relevant, nor how to resolve trade-offs, it just "leaves you to muck about on these questions as best as you can" (Foley 1993, 128). On the other hand, there are also worries that RE is *too demanding* to be applicable. Because of its coherentist and holistic character, it might be unworkable for imperfect epistemic agents like us (cf. Van der Burg and Van Willigenburg 1998, 4):

> [The] wide-ranging objectives of even a weak wide reflective equilibrium are at minimum intimidating and may be unattainable ideals of both comprehensiveness and coherence. (Beauchamp and Childress 2013, 405)

As a personal report, I can certainly confirm that applying RE can feel intimidating, and may sometimes involve some "mucking about". However, the case study demonstrates that reflective equilibrium is a powerful methodological framework, which puts real constraints on the justification process and provides helpful guidance. In this chapter, I elaborate these points, work out further implications, and show how, based on the case study and its results, these two fundamental worries can be addressed. The discussion will stay on a rather general level, as results of the case study both for precautionary principles as well as for reflective equilibrium have already been summarized and discussed in detail at the end of each phase in Chaps. 6–8.

T. Rechnitzer, *Applying Reflective Equilibrium*, Logic, Argumentation & Reasoning 27, https://doi.org/10.1007/978-3-031-04333-8_9

An important part of evaluating a method is asking whether it was suitable to pursue a given research goal—in this case, the justification of an action-guiding, moral precautionary principle. Thus, I start by discussing the main results for precautionary principles, and show how using RE contributed to them (Sect. 9.1).

I then focus on interesting results of the application of my specified RE method (in Sect. 9.2), before discussing more broadly what follows for RE as a method in philosophy (in Sect. 9.3). The case study shows that it is possible to specify RE as a method. However, I argue that it might be more fruitful to think of reflective equilibrium in the first place as a *methodology*, that is, a framework that guides decisions at various stages during the research process—including the selection of adequate methods. This also allows us to give a more satisfying answer to the two fundamental worries that RE is either too vacuous or too demanding.

9.1 Results of the Case Study for Precautionary Principles

The goal of this book was to explore what it would mean to seriously apply RE *as a method*, and to test whether this is both possible and fruitful. In the introduction, I distinguished between methods, methodology, and epistemology. I proposed to understand *methods* as concrete tools and techniques of research, in the sense of a set of instructions or steps which should be followed to achieve a given objective.

Consequently, whether the application of RE as a method was successful also depends on how well it was able to contribute to its objective. For the case study, the pragmatic-epistemic objective was to justify an action-guiding moral principle that is applicable to the subject matter of precaution and precautionary decision-making.

The results for precautionary principles are already discussed in detail at the end of each phase of the case study (Chaps. 6–8). This section therefore only provides a rough summary of how the Tripartite Precautionary Approach to Threats of Rights Violations (Rights-TPA) resulted from the application of RE, before outlining some possible next steps in the debate about PPs.

The Rights-TPA avoids the main objections PPs face, and is able to address the desiderata and open questions identified in Chap. 4. RE could significantly contribute to the formulation and justification of the Rights-TPA, but these are not the only benefits: the results of its application also allow us to gain further insights for the debate about PPs, and to identify possible next steps of research.

The **Tripartite Precautionary Approach to Threats of Rights Violations (Rights-TPA)**:

- The **Rights Meta Precautionary Principle (Rights-MPP)**: Uncertainty must not be a reason for inaction when there are threats of rights implications.
- The **Precautionary Tripod**: The elements that have to be specified in order to obtain an action-guiding precautionary principle version: If there is a threat that meets the *harm condition* (i.e., a specific rights violation) under a given *knowledge condition* then a *recommended precaution* should be taken.

- **Proportionality**: Demands that the elements of the Precautionary Tripod are adjusted proportionally to each other, understood as *Consistency*: The recommended precaution must not be recommended against by the same PP version, and *Efficiency*: Among those precautionary measures that can be consistently recommended by a PP version, the least costly one should be chosen.
  The **starting point for a rights-based PP version**: If there is (1) a threat of a wrongful rights violation, then (2) select the least stringent knowledge condition that results in a consistently applicable version of PP given the harm condition. To comply with the Rights-MPP, uncertainty must neither render the PP version inapplicable nor lead to continual delay in taking measures to prevent rights violations.

The Rights-TPA is what I call an integrated PP interpretation, that is, it combines action-guiding, epistemic, and procedural elements (Steel 2015, 14). The procedural element, the Rights-MPP, puts constraints on the selection of adequate decision-rules: decisions must be made in a way that ensures that uncertainty about the likelihood or harmfulness of threats of rights implications will not lead to inaction. This means, for example, that we need effective ways to make decisions about protective measures also in situations in which no reliable probability information is available.

As Steel (2015, 18) argues, the MPP is a substantive and informative principle: for example, it will recommend against the use of decision rules that allow precautionary action only if it can be shown that the expected benefits of the precaution outweigh its expected costs. Such a rule would lead to inaction due to uncertainty when expected costs and benefits cannot be reliably forecast (Steel 2015, 21). Consequently, it will often speak against the use of standard approaches of risk management, like cost-benefit analysis.

The Rights-TPA thereby avoids two of the main objections PPs face: it is neither vacuous nor redundant. Due to the Proportionality-Element, it also avoids the other main objections, i.e., that PPs are incoherent and can lead to paralysis, and that they are irrational because they allow merely speculative harm to be a reason for strict regulations that might do more harm than the original threat (see Chap. 4 for the main objections against PPs). The Precautionary Tripod has to be specified on a case-by-case basis in a way that meets the constraints of both the MPP and the Proportionality-element. This means that measures have to be selected that do not themselves meet the knowledge and/or harm condition of the PP version, blocking both (a) the possibility that measures are taken that pose a greater threat than the one they are supposed to prevent, as well as (b) the possibility that a PP version leads to paralysis by justifying a measure while at the same time recommending against the same measure. Additionally, the Rights-TPA has the benefit that it is not tied to a specific category of uncertainty, but can be adjusted with respect to varying degrees of knowledge and available evidence. This ensures that all available evidence will be taken into account, instead of, e.g., solely comparing possible worst cases. In the RE process, this characteristic contributed to the superiority of the Rights-TPA to the Rights-Maximin-PP, leading to its adoption as the current system at step A_6. For example, the Rights-TPA can account for more commitments than the Rights-Maximin-PP, and it does not require an additional criterion for "reasonable outcomes". Which outcomes should be included, i.e., which threats should be treated as "realistic enough", is built into the Proportionality criterion of the TPA: "reasonable outcomes" are those against which we still can take precautionary measures that do not themselves meet the harm condition and the knowledge condition of the threat.

The Rights-TPA is an adapted version of Steel's (2015) proposal for a PP. The main difference from Steel's proposal is the special attention paid to the protection of rights. This modification is one of the results of the RE process: at step A_4 during the process of adjustments, it became apparent that none of the current candidate systems could account for a subset of the current commitments that could all be interpreted as distinctly *moral* value commitments. These concerned, e.g., the difference between risk-taking and risk-imposing, the protection of human health and the environment, the protection of rights of future generations, or paying attention to those that would be worst off if an uncertain harm should materialize.

One possibility would have been to exclude these commitments from the subject matter, e.g., by arguing that the target PP should only apply to decisions where the relevant values are determined in some other way. However, as the pragmatic-epistemic objective of the case study was explicitly to justify a *moral* principle, such a move would have meant abandoning the initial objective. Additionally, value-commitments such as that human health or the environment deserve special protection are arguably central to the subject matter of PPs (see Chap. 4). It is thus at least plausible to construct a PP that expresses substantial values instead of being a purely prudential or rational principle that tells us which means to take to achieve a given end. As argued in phase 3 of the case study, adopting the protection of rights as the normative basis for a precautionary principle provides a unifying rationale, since most if not all relevant cases of harm to the environment or to human health will be subsumable under it, as will be cases of threat of serious or catastrophic harm.

As I argue in Chap. 8, the fact that the Rights-TPA turned out to be the most convincing candidate also supports the original approach of Steel (2015), which is shown to be a very comprehensive and systematic formulation of a precautionary principle. Adapting Steel's proposal to apply in particular to threats of rights violations makes it more substantial as a moral principle, which fits the pragmatic-epistemic objective of the case study. However, Steel can be interpreted to have the objective of formulating and defending a principle for public policy-making, and to refrain from making a commitment to substantial moral values or theories for this reason. Nonetheless, the Rights-TPA and the TPA are compatible to a certain extent: they will only lead to conflicting verdicts if the harm-condition of a PP-version of the TPA is set in a way that conflicts with the Rights-Meta-PP. Being clear about their relative objectives helps to see how they can both be *precautionary* principles, yet systematize different sets of resulting commitments.

This insight can be extended to the debate about precautionary principles more broadly. Talk about "the precautionary principle" without further qualification should be abandoned, e.g., statements about what "the PP" says, entails, or demands. Instead, it will be more fruitful to talk either about ideas commonly associated with PPs—that is, which can legitimately be seen as central commitments about the subject matter—or to refer to specific PP proposals that are defended with respect to a specific pragmatic-epistemic objective.

There might emerge one proposal that becomes authoritative at least for a specific context, making reference to "the PP" sensible in this context. However, at least given how diverse and at times fragmented the debate currently is, it is more fruitful to be as clear and explicit as possible.

Avenues for Further Development of PPs This leads us to the question of what we can learn from the results of the case study for further research on PPs. As we have seen, the Tripartite Precautionary Approach to Threats of Rights Violations (Rights-TPA) is a candidate for a moral PP which can be defended based on the RE criteria: it answers common objections, can account for a broad range of commitments, and does justice to theoretical virtues.

However, this does not mean that the debate is settled now once and for all. The account of justification that RE provides is holistic, but we cannot consider and evaluate everything at once. One consequence of this is that we typically have to work with working hypotheses and assumptions that we need in order to "scaffold" other parts of the position, before we can go back and re-examine and elaborate them (which then, in turn, can of course have further implications for the rest of the position) (cf. Elgin 1996, 106; 2017, 20). Next steps with respect to the Rights-TPA would be to explicate the concept of "uncertainty", and to elaborate the theory of rights that has to be assumed in the background in order to make the Rights-TPA applicable.

Another consequence is that it is always possible for further relevant considerations to emerge, which so far have been overlooked, and which unbalance the resulting position. In fact, this happened several times during the application of the RE method in the case study. This means that even a relatively stable position with a high degree of justification should never be seen as a final end point. Instead, I argue that each application of the resulting system is best conceived as an ongoing process of further elaborating, testing, and refining the position. The (Rights-)TPA has been shown to be justified to a high degree, but there is always the possibility that it leads to a verdict that we just are not willing to accept, i.e., which conflicts with a central commitment. Nonetheless, just because its justification is defeasible does not mean it is unreasonable to accept the Rights-TPA and use it to decide how we should proceed when facing threats of rights implications (see Chap. 2). At least as long as there is no other candidate that can be shown to be justified to a higher degree, we should commit to the Rights-TPA.[1]

[1] Or, if one does not want to commit to a moral PP, the TPA. I will speak of the (Rights-)TPA in the following to leave it open which candidate one does want to accept.

A useful way to work forward from here would be to apply the (Rights-)TPA systematically in applied ethics and case studies on policy recommendations. This would, on the one hand, mean that we can profit from the guidance of the (Rights-)TPA. On the other hand, this should be done in the spirit of RE, that is, we should see each such application also as an opportunity to further test and refine the system, and to search for potential problems and conflicts.

9.2 Results of the Case Study for the Method of Reflective Equilibrium

In this section, I focus on interesting results of the application of the RE method as it was specified in Chaps. 3 and 5. The next section will discuss more broadly what follows for RE as a method—or methodology—in philosophy. I start by summarizing how RE was specified as a method for the application in the case study, before highlighting interesting and relevant results from the application of this method.

9.2.1 Specifying an RE Method

In the first chapter, I introduced the distinction between epistemology, methodology, and specific methods. I roughly defined *epistemology* as a theory and analysis of what has epistemic value, *methodology* as a theory and analysis of how research should proceed, and *methods* as concrete techniques for conducting research (cf. Ackerly and True 2013).

Applying these distinctions to the present book, we can say that Chaps. 2 and 3 developed reflective equilibrium as a methodology which is based on an imperfect procedural epistemology that is weakly foundationalist and takes understanding as the goal of inquiry. Chapter 2 focused on the epistemological foundations of this methodology, discussing the conditions under which an epistemic position is justified. Chapter 3 focused on how this methodology can guide actual research: it highlights the challenges that need to be addressed, and the decisions that need to be made, if one wants to specify a method of reflective equilibrium that can be applied. The chapter provides guidelines for how one can proceed in order to obtain an applicable method, and identifies steps that will help to structure the research process. This is in line with the idea that a methodology guides decisions at various stages during the research process, e.g., which methods are chosen, and provides a defense of those decisions to one's academic peers (cf. Ackerly and True 2013, 137).

Chapter 5 then demonstrated how a workable, specified method of reflective equilibrium can be obtained by addressing the challenges and decisions outlined by the so-described RE methodology. In particular, this required that we concretize the two steps of an RE process of alternatingly adjusting commitments and adjusting the system. This was done through specifying the RE criteria, e.g., spelling out "agreement between commitments and system" in the form of an account-function.

Importantly, describing the starting position of the RE application also included clarifying my pragmatic-epistemic objective for the RE application. As explained in Chaps. 2 and 3, how the RE criteria are specified and weighted also depends on the particular objective of the process of inquiry. In my case, I did set the pragmatic-epistemic objective as "Justifying an action-guiding moral principle that is applicable to the subject matter of precaution and precautionary decision-making". This subject matter has been described in Chap. 4, which gave a survey about different interpretations of precautionary principles (PPs), along with justifications that have been brought forward for them as well as objections against them. Chapter 4 thereby also provided the basis for identifying relevant commitments, background elements, and candidates for the system.

Notably, the RE methodology as described in Chaps. 2 and 3 turned out to be elaborate enough to allow me to concretize RE as a method and to identify its input. This is not to say that this is the only way to specify RE as a method. In particular, I did work with some approximations and simplifications, where others might want to use more refined measures. For example, I only assigned rough ordinal weights of low–medium–high to commitments, but then *de facto* measured these weights on an interval scale when defining my account function in order to allow me to get an approximate comparison of how well different candidate systems can account for current commitments. There is no reason to think that this is the uniquely best, or even just one of the best, ways to measure these criteria. However, it shows that it is possible to operationalize the RE criteria in an applicable way, and thus lends support to the developed methodology. Further research and applications can now refine the ways in which RE can be concretized as a method, and draw on various epistemological theories to develop ways in which criteria like independent credibility of commitments or support from background theories can be spelled out more precisely.

9.2.2 Results from the Process of Adjustments

Once RE was specified as a method, the two steps of either adjusting commitments or adjusting the system could be applied to structure the process of adjustments. This application could, on the one hand, vindicate and illustrate certain aspects of the epistemological conception of RE. On the other hand, it also had some unexpected yet insightful results that help to further develop our understanding of RE.

Applying RE Can Fruitfully Guide the Development of Epistemic Positions As in particular the first phase of the case study shows, RE is not only a method to achieve balance between considerations that are all already made explicit and developed. For example, I started by assessing two of the commitments—the Rio

PP and the Wingspread PP—as candidates for principles, but they did not meet the RE criteria to a sufficient degree and were rejected. However, identifying their shortcomings allowed me to formulate *guiding questions* for the systematic exploration of commitments, and for developing improved candidate systems. Additionally, this illustrates that fleshing out the set of commitments, and filling in gaps, can also be an important result of an RE process—not only adjusting commitments under pressure from an existing, fully developed system, but also when searching for answers to open questions. This brings us to another insight, namely the important role of emerging commitments.

Emerging Commitments Are an Important Aspect of RE Any application of RE as a method obviously depends partly on the input that we explicitly consider. As explained in Chap. 3, we typically do not have a complete overview of all our commitments, meaning that we have to work with a relevant selection,[2] and that we should continuously be on the lookout for further relevant commitments. Such emerging commitments, even though they were not initially explicitly considered, still count as input commitments, that is, they constrain the subject matter and have a degree of credibility that is independent of their agreement with the current system.

Throughout the process of applying the RE steps, emerging commitments turned out to play an important and sometimes decisive role. For example, when searching for further relevant commitments during step B_5, several commitments emerged that caused problems for the Rights-Maximin-PP. In particular, as soon as the conditions of the Rights-Maximin-PP are met, it recommends that we select the course of action with the best worst case, disregarding additional information that we have, e.g., on the likelihood of the possible outcomes. However, based on arguments from the literature, I adopted the commitments EC 28 and EC 29 at step B_5:

EC 28 A decision principle for decisions under uncertainty needs criteria to decide which outcomes should still be included as "reasonable" or "plausible" enough. [medium] [emerged at Step B_5]

EC 29 The information we have about possible outcomes of courses of actions should not be irrelevant for the decision process only because it is not sufficient to assign reliable probabilities. [medium] [emerged at Step B_5]

Both of these commitments are in conflict with the Rights-Maximin-PP, which led me to introduce the Tripartite Precautionary Approach to Threats of Rights Violations (Rights-TPA). The Rights-TPA can account for more commitments than the Rights-Maximin-PP, and it does not require an additional criterion for "reasonable outcomes". Which outcomes should be included is built into the Proportionality criterion of the TPA: "reasonable outcomes" are those against which we still can take precautionary measures that do not themselves meet the harm and knowledge conditions of the threat.

[2] I describe my own selection criteria in Chap. 5.

A Dynamic and Non-Linear Process That the Rights-TPA is chosen over the Rights-Maximin-PP at step A_6 illustrates how in the course of a process, the acceptability of a candidate can change: at step A_4, the TPA was rejected in favor of the Maximin-PP. At step A_6, further information was made explicit, turning the scales in favor of the TPA.

This leads us to the more general point that the RE process is dynamic and non-linear. Firstly, previously rejected elements, like (parts of) a candidate system or specific commitments, can become acceptable later on, and vice versa. Secondly, re-adjustment of commitments is possible, meaning that commitments that were already adjusted can be re-adjusted (for example, at Step B_6, C 13 replaced C 4 which had previously replaced the initial commitment IC 12). Thirdly, the set of explicitly considered commitments can change throughout the process. I have already stressed the importance of emerging input commitments. Another way in which the set of current commitments might be expanded is through inferences from the current system, like the newly inferred commitments in step B_2. And fourthly, it is possible that parts of the background should move to the foreground and vice versa. For example, in phase 2, the explication of "Precautionary Measures" from phase 1, *ExplicPrec*, lost its relevance for the position in the foreground and could be moved to the background: the relevant candidate systems at this point, the RCPP and the Maximin-PP, will both only select measures that fulfill the criteria for being a precautionary measure, without needing *ExplicPrec* to yield a verdict.

"Sub-processes" are Possible, in which Only a Part of the Position is Adjusted The explication of concepts, like "precautionary measures" in Chap. 6, or developing parts of the system, like the rights threshold in Chap. 8, are examples of "sub-processes" within the RE process. The case study shows that such sub-processes, in which only a part of the system and a subset of the commitments are adjusted with respect to each other, can be integrated into the RE process.

Relevance of the Background for the Position in the Foreground What can or cannot reasonably be seen as part of the background has significant implications for the interpretation and adjustment of elements of the position in the foreground. Here are three especially important ways in which this can be the case: Firstly, whether or not a system can account for a commitment will often depend on the available background information, e.g., the factual information about a case. Secondly, whether or not a system has a theoretical virtue can also to some degree depend on background elements. See the example of the virtue of determinacy in Chap. 6: here, candidate systems have a low determinacy because terms that they use are not clearly defined in the background. Thirdly, how much weight should be given to a commitment will also partly depend on whether or not it can be supported with relevant background theories.

Functional Difference between Commitments and System I argued that the relevant difference for RE is not between particular judgments and general principles, but between the attitude of commitment on the one side, and the ability to provide a systematic account on the other (see Chap. 2). There are different constraints on both sides, which is also specified in the steps for adjusting the system (A-Steps) and the steps for adjusting commitments (B-Steps) of the RE method. While a system has to be able to account for commitments while doing justice to theoretical virtues, commitments have to be in agreement with the system, have to respect input commitments, and must have some independent credibility. But none of this requires that commitments and a system have to be different from each other in content or in form.

The application of the RE method shows that this is a plausible way to draw the distinction. For example, in phase 1, the same propositions could be used both as commitments and as candidate systems. In other cases, whole classes, or subsets, of commitments were discussed and adjusted, e.g., whether all commitments referring to situations in which probabilities are available should be excluded from the subject matter. This shows that not only commitments to particular judgments are relevant, but also—and maybe even more so—general commitments.

Additionally, some ways in which a system fails to account for commitments can be especially insightful, e.g., it can become apparent that they all have something in common. For example, when adjusting commitments with respect to the Maximin-PP in step B_4, one thing that became apparent was that the Maximin-PP could not account for a subset of the commitments that could all be interpreted as distinctly *moral* commitments. As the pragmatic-epistemic objective (see Chap. 5) was explicitly to formulate a *moral* principle, this is problematic for the Maximin-PP. However, the other candidates for the system up to this point faced the same problem. This issue was addressed in phase 3 through a sub-process in which the rights threshold principle was developed to supply the Maximin-PP as a part of the system.

Reasonable Constraints, but No Rigid Rules Let me finish with a more general result on the kind of guidance that the RE method provides. While the method did not provide rigid rules that unequivocally determined which adjustments should be made, it proved to be applicable in an informative way that put real constraints on the process of justification. This is in line with how the epistemology and methodology of RE were described in Chaps. 2 and 3: trade-offs can be resolved in different ways, and there is no reason to think that there will always be a single best way to adjust a position. RE allows for reasonable pluralism, and a specification that does not include this would not be adequate.

That there was not always a single, unequivocally best way to adjust the position does not mean that the adjustments were arbitrary, either. Using the RE criteria to assess and compare candidate systems in the A-steps, or to assess commitments and different ways to adjust them in the B-steps, provided the basis for an informed choice, and to bring forward arguments in favor or against different ways to move forward. For example, comparatively assessing the Principle 3 System, the Rawlsian Core Precautionary Principle (RCPP), and the Utilitarian Uncertainty Principle (UUP) in phase 2 allowed me to clearly describe their respective strengths and weaknesses, and to make trade-offs explicit. On this basis, I could bring forward an explicit argument for adopting the RCPP.

Similarly, when adjusting the current commitments with respect to the RCPP at step B_3, the RE criteria allowed me to identify which commitments are in conflict, and to assess how different ways to adjust them would influence the position. Commitments were not blindly adjusted in order to increase agreement with the system, but their independent credibility, expressed in form of the low–medium–high weights, had to be respected. For example, the following commitment:

EC 15 Precautionary measures should not introduce serious threats of their own. [low] [emerged at Step B_2]

Was replaced by the new commitment:

C 3 Precautionary measures should not introduce threats that are equally or more severe than the threats they are aimed at, i.e., threats that have the same or a greater potential for harm. [medium] [replaced EC 15 at Step B_3]

The reason for doing so was that this was a defensible way to increase agreement with the system. To avoid paralysis in cases where there is no completely safe option, one should not demand that precautionary measures never introduce any serious threat of their own. Arguably, the important point is that the overall threat level is reduced, making it reasonable to replace EC 13 through C 3.

However, there were no convincing arguments why cases where probabilities are available should be excluded from the subject matter, even though the current system at step B_3—the RCPP—could not account for those commitments. Consequently, commitments referring to cases with available probabilities were not rejected, even though it would have increased agreement between the commitments and the system.

This exemplifies how applying the RE steps allows us to defend adjusting commitments and selecting candidate systems based on arguments. Using the RE criteria to assess and compare candidate systems, or to assess commitments and different ways to adjust them, provides the basis for an informed choice based on arguments in favor or against different ways to move forward. At the same time, these adjustments and the reasons behind them can be presented in a way that is traceable by others, which would allow them to go back in the process and to explore what outcomes one would reach if one were to make different choices.

9.3 Discussion: Reflective Equilibrium as a Methodology and Method in Philosophy

In the first chapter, we identified a range of challenges for applying RE as a method, most of them having to do with the fact that it was unclear what it would entail to apply it, e.g., what its input would be, how we should proceed when making adjustments, or how we could assess whether or not an actual state of RE was reached. Previous attempts at applying RE have not addressed these challenges to a satisfying degree. The case study in this book shows how they can be addressed and how an applicable method of RE can be obtained and put into practice. On this basis, we can now address the two fundamental worries that concern whether RE is either too vacuous, or too demanding, to be an applicable method.

The focus of this book was on applying reflective equilibrium *as a method*. In this last section, I want to take a step back in order to adopt a broader perspective. We saw that RE can be specified and applied as a method. But is this the most fruitful way to think about it and to make use of it? That is, if we draw the distinction between methodology and method in the way I suggested, does it make the most sense to see RE as a method in the sense of a set of instructions that should be followed in order to achieve a given objective? I argue that it might be more fruitful to develop and use RE as a *methodology* in philosophy—and that this will allow us to give a more convincing answer to the two worries.

I start by reconsidering some aspects of the case study that might cast doubt on whether RE is best understood as a method. The case study was specifically designed as a test for a step-by-step, open-ended application of the method of reflective equilibrium. In the form of the A- and B-steps and the stopping rule, the method provides guidelines that should be followed in a particular order. And Chaps. 6–8 describe a process of adjustments that develops a position through following these steps.

However, to what degree do these chapters describe what I actually did when constructing my position? That is, does the case study show that the RE method provides us with a set of explicit instructions that we can and should follow when constructing an epistemic position? Does it prescribe the exact steps we should take when we want to formulate and justify, e.g., a theory that does not yet exist?

In fact, the presentation of the process of adjustments in this book is to a significant degree a "cleaned up" reconstruction. For example, choices had to be made how to divide the process into three phases in order to be able to demonstrate and exemplify different aspects. Some of the confusing and ultimately misleading pathways have been left out in the final presentation of the process in order to achieve at least some comprehensibility for the reader. But even in this reconstruction, not everything strictly follows the logic of the A- and B-steps. Think about the sub-processes, like the explication in phase 1, or the formulation of guiding questions: while these could be integrated, and are interesting results, they also show that it was sometimes necessary to "tweak" the steps a bit to keep following the structure of the two alternating steps of going back and forth between commitments and a system. This "tweaking" does not directly speak against the possibility of spelling out an RE method—in fact, further developments of the method could try to incorporate this and allow for a more flexible structure. However, more generally, the dynamic and non-linear structure of the RE process casts doubt on the degree to which we can have a set of instructions that should be followed in a given order. Even though the case study shows that adjustments were not arbitrary, and RE provides helpful guidelines, this might not be enough to satisfy critics who see RE as too vacuous and permissive to count as a real method. They could argue that, after all, a lot depends on other philosophical methods that are applied as part of the RE applications, like deductive arguments, explications, or thought experiments (e.g., to make commitments explicit).

Once we take a step back from the case study, we see that it might be more fruitful to think of RE as a methodology, and not as a set of specific instructions that should be followed in a particular way. Let me elaborate.

Process Versus State of Reflective Equilibrium If we understand RE primarily as a methodology, this raises questions about the role of the process of adjustments. If the process is not primarily understood in terms of describing a method by providing steps that should be followed, then what is its role? Why do we even need the process in the first place—would it not be enough to specify the conditions of having reached a state of reflective equilibrium? The important thing seems to be the resulting position and whether it can be defended. How we get there could be seen as a part of the process of discovery and irrelevant for justification. However, there are several reasons why we still should take the process seriously.

First, there are epistemological arguments for the process: it is correct that we should not conflate our psychological processes with our epistemic justification. The process of adjustments as part of RE is not intended as a description or prescription of actual cognitive practice (Baumberger and Brun 2021, 7936), but as a reconstruction. We do not have to recount in detail the genealogy of our position.

However, we need to be able to reconstruct a process of adjustments from the input to the resulting position in order to assess whether the resulting position is justified. As Baumberger and Brun (2021) argue, reconstructing such a process will often be the only way to decide whether a position is in reflective equilibrium. I see two main reasons for this: one is that the clarification and specification of the relevant configuration of epistemic goals is part of the process of equilibration (see also Chap. 3).

> We may know, for example, that our pragmatic-epistemic objective demands that we do not give up too much precision in favor of simplicity, but how much we must finally give up is just as much a research question as the question of which commitments we will end up with in reflective equilibrium. (Baumberger and Brun 2021, 7937)

Another reason for not cutting the process out of the account of justification is that an RE state is not something that can be reached as a final product that is guaranteed to stay stable (cf. Bonevac 2004). We cannot completely survey all of our commitments, and it is always possible that new considerations should arise that unbalance the position, or that something changes in the background. Instead, we have a continuous progression of positions that are more or less justified, more or less stable, and connected through processes of adjustments. This does not render the RE state insignificant, however. It is still important as an ideal because it articulates what we are searching for. We can assess and analyze positions with respect to this ideal, thereby informing how the process can be continued.

That we can analyze positions with respect to this ideal does not mean that we know beforehand what the ideal position would be, otherwise we could simply adopt it. Instead, it means that the RE criteria can help us to identify potential weaknesses of our positions, i.e., where they are lacking with respect to an ideal RE state. In turn, this helps us to defend them, if, e.g., at the moment we cannot identify any problems.

Second, even if the process-aspect of RE is intended as a rational reconstruction and not as a description or prescription of how inquiry should proceed, it still is a powerful *methodological* framework. It will often be worthwhile to conduct one's research along the lines of RE, e.g., to try to be explicit about what the input commitments are, to explicitly state the pragmatic-epistemic objective one pursues and what implications this might have for, e.g., the theoretical virtues one expects one's system to have, and to document how one's position changes. This provides useful heuristics and guidelines without forcing one's thinking into a strict corset. At the same time, it will better enable one to later show via a reconstruction as an RE process that one's resulting position can be defended as being in a state of reflective equilibrium. Discussing and defending results in this way will also help to identify gaps and avenues for further research, for example, which concepts still need to be explicated, identifying potential tensions between commitments, or between the position and background theories, etc.

Uses and Benefits of RE as a Methodology Reflective equilibrium can provide a fruitful methodological framework for philosophical inquiry in different ways: it can be used as methodological approach to structure one's pragmatic-epistemic projects, or to reconstructively appraise positions, but also to analyze debates and to compare positions. For example, one can analyze a debate in terms of different agents trying to bring more-or-less overlapping sets of commitments into different RE states, with potentially differing pragmatic-epistemic objectives. Such an analysis provides the basis for situating oneself in the debate, and to identify problems worth addressing.

Using the RE framework in this way would also help to connect seemingly isolated, small-scale projects to bigger epistemic projects and allows us to structure the academic division of labor. Each RE process, even if the RE conception is designed for a single epistemic agent, ultimately has to be seen as part of a collective effort.

One reason is that each RE process of developing a position in the foreground has to take place against a background of other theories, factual knowledge, and assumptions. Parts of this background, at least if they are supposed to support the position in the foreground, have to be defended based on their own RE processes, and those will often be conducted by other agents. Another reason is that it is desirable to develop many different positions, and to follow even those pathways that might not initially seem especially promising. They might turn out more successful later than initially anticipated. But, importantly, even if they turn out to be unfruitful, we have learned something more, and are better able to defend alternative positions that can be developed.

Understood as a methodology, RE does not prescribe particular methods or exact criteria no matter the question (cf. Ackerly and True 2013, 137). This fits with Walden's (2013) defense of RE as non-essentialist, rejecting the idea that any kind of inputs or specific standards should be regarded as fixed. However, I argue that this characterization is much better suited for a methodology than for a method: a method that does not define any kind of specific standards could indeed rightly be accused of being vacuous and not helpful. A methodology, however, provides the framework to select appropriate methods for pursuing a given research goal, that is, for specifying criteria, for identifying and selecting input, for interpreting results, and so on.

Going beyond general appeals to RE and instead using it explicitly as a methodology promises to be highly beneficial for philosophical inquiry and debate. It not only provides useful guidance, but also forces us to make things explicit in a way that are specific advantages of RE. Let me name three examples. Firstly, it is important to acknowledge the functional difference between commitments and a system. They both have to meet different constraints, which we can only assess when we are clear about the respective roles. For example, assessments of theoretical virtues like simplicity only make sense for parts of a system, whereas asking whether something has independent credibility—e.g., through being intuitive—only makes sense if we are talking about commitments.

Secondly, differentiating between input and resulting commitments allows us to distinguish between a pre-systematic conception of a subject matter and a systematic account of it, which organizes and re-interprets parts of the subject matter in a new way. As the example of precautionary principles shows, this helps us to draw a distinction between commitments on precaution and precautionary principles that might be inconsistent, in tension, or simply unconnected to each other on the one hand, and developing a systematic account of this subject matter which might differ substantially from the initial conception.

Thirdly, the distinction between background and foreground forces us to make as explicit as possible what we presuppose. It also enables us to situate individual epistemic projects in a bigger context. By encouraging us to flesh out our positions, e.g., by systematically searching for further relevant commitments, applying RE will contribute to our understanding—even in situations where nothing is adjusted.

A Place for RE *Methods*? I have argued that, ultimately, adopting reflective equilibrium as a methodology is more fruitful than to keep talking about it as a method. When adopting the RE methodology, some methods will be more appropriate for certain questions and subject matters than others: e.g., in political philosophy or applied ethics, working with questionnaires and taking into account folk commitments might be more appropriate than in, e.g., philosophy of science. A next step in research on RE could thus be to think about this more explicitly.

Also, I do not want to exclude that it is possible and useful in some sense to spell out an RE method. There might be certain areas, e.g., applied ethics or medical ethics, where it is possible to spell out an RE method to address particular problems. Especially in situations in which the RE process is relatively narrow and standardized, it seems likely that such a method could be specified (cf. Van der Burg and Van Willigenburg 1998, 13–15). What I here mean by "narrow" are cases in which there is, e.g., broad agreement on relevant background elements, and only a small part of the position in the foreground needs to be adjusted, which might be restricted to finding a solution to one specific problem—like selecting an appropriate treatment for a patient.

Summing Up and Answering the Two Fundamental Worries I have argued that there is not a single, specific *method* of reflective equilibrium, consisting of specified standards on what kind of input to include, or rigid steps that should be followed. Instead, I argued that reflective equilibrium can be understood as a *methodology* which is based on an imperfect procedural epistemology that is weakly foundationalist. From this, it follows that objections to particular ways to spell out RE as a method—e.g., as taking intuitions as its input—do not necessarily amount to objections to the methodology more broadly. RE *methods* are specified with respect to a particular purpose, and in the spirit of the RE epistemology, are themselves revisable (Elgin 1996, 12). In a sense, reflective equilibrium is not a method among others; even specified RE methods might contain other things that are often counted as methods, like deductive inference, explications, or inductive arguments.

As an answer to the worry that RE is too vacuous to be a method, and at best can provide "meta-advice", we can say that reflective equilibrium as a methodology is no more "meta" than other methodologies: it tells us how to approach processes of inquiry, and gives us guidelines for how to make decisions on what methods will be appropriate to address a specific problem of justification. In fact, it is an important element of RE that it does not impose rigid standards of justification, and takes seriously that we already have normative standards and theories that we have reason to regard as justified, and that should inform the ways in which RE is spelled out and applied.

As an answer to the worry that RE is too demanding, and will be overwhelming for imperfect epistemic agents, we can say the following: it is true that RE demands that we take many different things into account and that we balance them against various constraints. This certainly is demanding. However, just because something is demanding does not mean that it is implausible—justification is no trivial task. And, importantly, justification in reflective equilibrium is inherently provisional and an ongoing process, which takes the edge off this concern. Instead of being overwhelming and paralyzing, RE shows us how we can move forward and identify what can or cannot be justified *at a given time*.

As an answer to both worries, I have shown in my case study how the RE methodology can be successfully specified in a way that provides constraints which are informative without being arbitrary.

References

Ackerly B, True J (2013) Methods and methodologies. In: Waylen G, Celis K, Kantola J, S Laurel W (eds) The Oxford Handbook of Gender and Politics. Oxford University Press, Oxford, pp 135–153

Baumberger C, Brun G (2021) Reflective equilibrium and understanding. Synthese 198(8):7923–7947. https://doi.org/10/ggkp4w

Beauchamp TL, Childress JF (2013) Principles of Biomedical Ethics, 7th edn. Oxford University Press, Oxford

Bonevac D (2004) Reflection without equilibrium. J Philos 101(7):363–388

Elgin CZ (1996) Considered judgment. Princeton University Press, Princeton

Elgin CZ (2017) True enough. The MIT Press, Cambridge, MA

Foley R (1993) Working without a net: A study of egocentric epistemology. Oxford University Press, New York

Steel D (2015) Philosophy and the precautionary principle. Cambridge University Press, Cambridge

Van der Burg W, Van Willigenburg T (1998) Introduction. In: Van der Burg W, Van Willigenburg T (eds) Reflective equilibrium: essays in honour of Robert Heeger, library of ethics and applied philosophy, vol 2. Springer, Netherlands, pp 1–25

Walden K (2013) In defense of reflective equilibrium. Philos. Stud. 166(2):243–256. https://doi.org/10.1007/s11098-012-0025-2

Appendix A
Elements of the Reflective Equilibrium Process

This appendix contains lists of different elements that are part of the application of reflective equilibrium in Chaps. 6–8: input commitments (Sect. A.1.1), new commitments that resulted from the process (Sect. A.2.2), candidate systems (Sect. A.3), and the case descriptions behind the toy examples in the background (Sect. A.4). The reasons behind selecting those elements are explained in Chap. 5, which describes the design of the case study.

All of the listed commitments, etc., were considered when conducting the process of adjustments, but, as explained in Chap. 5, only important or exemplary aspects could be discussed in detail throughout the presentation of the application in Chaps. 6–8.

A.1 Input Commitments

Input commitments are those commitments that enter the process because we are committed to them independently of the process of adjustments. That is to say, a commitment is an input commitment either because we hold it initially, before the process starts (these are the initial commitments, see Sect. A.1.1), or because it becomes explicit during the process, but not by being inferred from the current system (these are the emerging commitments, see Sect. A.1.2). For more information on this distinction, see Chaps. 2 and 3.

Commitments have a weight of either low, medium, or high assigned to them. These weights serve as a rough indication of the (independent) credibility of a commitment. For more on independent credibility, see Chaps. 2 and 3. For more on how I assigned and handled those weights, see Chap. 5.

T. Rechnitzer, *Applying Reflective Equilibrium*, Logic, Argumentation & Reasoning 27, https://doi.org/10.1007/978-3-031-04333-8

A.1.1 *Initial Commitments*

As explained in Chap. 5, I selected my initial commitments for the three groups of (1) general commitments about precaution and precautionary principles, (2) commitments to judgments in simplified "toy examples", and (3) commitments concerning research and development on solar radiation management (SRM) as an exemplary real-world case where precaution is relevant.

A.1.1.1 General Commitments About Precaution and Precautionary Principles

IC 1 Where there are threats of serious or irreversible damage, lack of full scientific certainty shall not be used as a reason for postponing cost-effective measures to prevent environmental degradation. (Principle 15 of the Rio Declaration) [low]

IC 2 When an activity raises threats of harm to human health or the environment, precautionary measures should be taken even if some cause-and-effect relationships are not fully established scientifically. (Wingspread Formulation of the Precautionary Principle) [low]

IC 3 *Pro tanto*, it is better to take precautionary measures now than to deal with serious harms to the environment or human health later on. [high]

IC 4 Threats to the environment or to human health warrant special precaution because such harm is especially prone to have long latent periods, and to be hard if not impossible to remediate or compensate. [medium]

IC 5 Don't risk great harm in pursuit of modest benefit. [high]

IC 6 If we are not sure whether a substance or technology is safe, but have a viable alternative that can be shown to be safe (at least with higher certainty than the option in question), we should use the alternative, even if it might be more costly in economic terms. [high]

IC 7 Morally, a higher degree of precaution is required when making decisions that will have effects on others: when making decisions that will only affect yourself, precaution is a question of rationality, depending on your preferences and beliefs; but when making decisions that threaten to harm others, precaution is morally required. [medium]

IC 8 The structure of a PP includes two "trigger conditions", threat and knowledge, and a precautionary response. [low]

A.1.1.2 Commitments About Toy Examples

IC 9 You should wear adequate protective clothing when making soap. [high]

IC 10 In case 10, *Worst Case being Shot*, you should choose option B. [high]

IC 11 In case 9, *Job Offers*, you should choose the job in Chicago. [medium]

IC 12 The claim that the measles/mumps/rubella (MMR) vaccine might cause autism is not a reason not to vaccinate your child. [medium]

IC 13 You should vaccinate your child with the measles/mumps/rubella (MMR) vaccine, even if some people claim that the vaccine might cause autism. (Example from Hansson (2016, 90–91).) [high]

IC 14 You find a firearm, and from examining it, you come to the conclusion that it is not loaded. But you are aware that you don't know much about weapons—this is, in fact, the first firearm you have ever held in your hands. You must not point it at someone else and pull the trigger. Neither should you do the same with yourself. [high]

IC 15 In case 5, *Asbestos 1*, we should choose option (iii), Research&Regulation. [medium]

IC 16 In case 6, Asbestos 2, we should choose option (iii), banning asbestos and substituting it with other, safer substances. [medium]

IC 17 In case 13, *Genetically Engineered Algae*, the technology of genetically engineering the microalgae should not be prohibited based on the uncertainty about potential dangers from the technology (example from Hansson 2016, 96)). [medium]

IC 18 In case 7, *Disproportionate Outcomes 1*, the option "Safe 1" should be chosen. [high]

IC 19 In case 7, *Disproportionate Outcomes 1*, the option "Safe 1" should be chosen **because** the worst case of option "Risky 1" is disproportionately worse than what we could gain from it as compared with "Safe 1". [medium]

IC 20 In case 8, *Disproportionate Outcomes 2*, the option "Safe 2" should be chosen. [high]

IC 21 In case 8, *Disproportionate Outcomes 2*, the option "Safe 2" should be chosen **because** the worst case of option "Risky 2" is disproportionately worse than what we could gain from it as compared with "Safe 2". [medium]

IC 22 In case 12, Chemical Waste, the company should not be allowed to discharge the chemical waste into the lake (example from Hansson 2016, 96). [high]

A.1.1.3 Commitments About R&D on Solar Radiation Management (SRM)

IC 23 Independently of whether or not SRM should be considered as part of precautionary measures in case the globally implemented mitigation and adaptation strategies turn out to be insufficient to prevent dangerous climate change, it should not be used as the only precautionary measure. ("SRM" here is short for "research and development on solar radiation management in order to have it ready to use should dangerous climate change be imminent".) [high]

IC 24 SRM should not be considered as the only precautionary measure against the threat that the globally implemented mitigation and adaptation strategies turn out to be insufficient to prevent dangerous climate change, because it is inadequate as a precautionary measure. It is inadequate because: it introduces threats of its own, it is uncertain whether it would work in the intended way without unforeseen (negative) side-effects, and it imposes costs and responsibilities (e.g., for maintenance) on future generations. [medium]

IC 25 Non-invasive research into SRM should be done, as long as this does not negatively interfere with the search for and discussion of other approaches. [medium]

IC 26 A necessary condition for any application of SRM against harmful impacts of climate change is that it has to be accompanied by a strict mitigation and adaptation program that would allow us to stop doing SRM again as soon as possible. [medium]

IC 27 A necessary condition for any application of SRM is that its international and intergenerational governance is clarified and established in a morally sound way. [medium]

IC 28 In case 2, *R&D into SRM*, we should choose option (i), implementing an R&D program with the objective of developing SRM ready to use. [high]

IC 29 In case 3, *R&D into SRM, two kinds of research*, we should choose option (ii), doing non-invasive research into SRM, especially the aspects that contribute to our general understanding of climate science. [medium]

IC 30 If a climate catastrophe, e.g., runaway climate change, is imminent or already under way, and we have assessed all our options, and only the deployment of SRM promises to provide a chance to at least alleviate the worst impacts, thereby buying us some time, then SRM may be deployed even if there are still uncertainties about its effectiveness and its side effects. [medium]

IC 31 Precautionary measures should be taken against the threat that the globally implemented mitigation and adaptation strategies turn out to be insufficient to prevent dangerous climate change. [high]

A.1.2 Emerging Commitments

EC 1 The target PP is neither restricted to threats to specific entities (e.g., the environment and/or human health), nor is there a category of threat that takes lexical priority for the application of a PP insofar as it is a threat to specific entities. [low] [emerged at Step B_1]

EC 2 Serious threats *pro tanto* warrant precaution. [low] [emerged at Step B_1]

EC 3 The seriousness of a threat is assessed according to (i) the potential for harm of the threat, and (ii) whether or not the possible harm is seen as reversible. [low] [emerged at Step B_1]

EC 4 All *plausible* serious threats *pro tanto* warrant precaution. [low] [emerged at Step B_1]

EC 15 Precautionary measures should not introduce serious threats of their own. [low] [emerged at Step B_2]

EC 16 The costs and responsibilities for precautionary measures should be distributed in a morally sound way. [high] [emerged at Step B_2]

EC 17 The price of a precautionary measure consists of—compared with the course of action entailing the threat it is supposed to address—foregone benefits,[1] foregone opportunities, and additional threats. [medium] [emerged at Step B_2]

EC 18 The price of precaution should be proportional to the seriousness of the threat, given the available alternatives. [low] [emerged at Step B_2]

EC 19 The price of precaution should be proportional to the seriousness and the plausibility of the threat, given the available alternatives. [low] [emerged at Step B_2]

EC 20 The target PP applies to plausible and serious threats and prescribes measures that are proportional to the severity and plausibility of the threat. [medium] [emerged at Step B_2]

EC 21 The wider the dispersal of the outcome value distribution of a course of action, the more precaution is warranted by its negative outcomes. [medium] [emerged at Step B_3]

EC 22 *Pro tanto*, cases that involve threats of disproportionate harm warrant precautionary measures, even if this harm is very unlikely. [medium] [emerged at Step B_3]

---

[1] I take it that "foregone benefits" also includes direct monetary costs of precautionary measures that are spent, e.g., on installing safety measures, since the money used there cannot be spent for other purposes.

EC 23 *Pro tanto*, threats of harm to human health have lexical priority for precaution. [low] [emerged at Step B_4]

EC 24 *Pro tanto*, threats to the environment have lexical priority for precaution. [low] [emerged at Step B_4]

EC 25 When evaluating possible outcomes of courses of actions, the rights of future generations must not be discounted. [low] [emerged at Step B_4]

EC 26 When taking precautionary measures against a threat, attention has to be paid to those who would be worst off if the harm should materialize. (Distributional concerns matter for precaution.) [high] [emerged at Step B_4]

EC 27 Serious threats that can be addressed by an earlier generation must not be deferred to future generations. [low] [emerged at Step B_4]

EC 28 A decision principle for decisions under uncertainty needs criteria to decide which outcomes should still be included as "reasonable" or "plausible" enough. [medium] [emerged at Step B_5]

EC 29 The information we have about possible outcomes of courses of actions should not be irrelevant for the decision process only because it is not sufficient to assign reliable probabilities. [medium] [emerged at Step B_5]

A.1.3 Emerging commitments on What Counts as "Precautionary Measures"

EC 5 To wear appropriate footwear when going on a hike is not a precautionary measure. [low] [emerged at Step B_1]

EC 6 Looking left and right before crossing the street is not a precautionary measure. [low] [emerged at Step B_1]

EC 7 To bring a parachute when planning to jump out of an airplane is not a precautionary measure. (Example from Sandin 2004) [medium] [emerged at Step B_1]

EC 8 To have a parachute on board when planning to fly somewhere is a precautionary measure. (Example from Sandin 2004) [medium] [emerged at Step B_1]

EC 9 You went to a costume party and brought a fire extinguisher as part of your costume, and then were able to extinguish a fire that for some reason broke out: Bringing the fire extinguisher was not a precaution because you did not bring the fire extinguisher with the intention to put out fires. [medium] (Example from Sandin 2004) [emerged at Step B_1]

EC 10 Making an effort to be cautious and focused when frying food in scorching hot oil is not a precautionary measure against burning yourself, because it is almost certain that if you are absent-minded while doing this, you will burn yourself at some point. [low] [emerged at Step B_1]

EC 11 Chewing your food is not a precautionary measure against choking. [medium] [emerged at Step B_1]

EC 12 As a factory worker who is well informed about the dangers of being exposed to the hazardous chemical X in your work, performing a ritualistic dance to protect you from a hazardous chemical is not a precautionary measure. (Example from Sandin 2004) [high] [emerged at Step B_1]

EC 14 Research and development (R&D) into solar radiation management (SRM) in order to have it ready to use should dangerous climate change be imminent, is not, on its own, a precautionary measure against the threat of dangerous climate change. [medium] [emerged at Step B_2]

EC 13 Precautionary measures should be effective in preventing or substantially ameliorating either a threat or the harm of a threat. [high] [emerged at Step B_1]

A.2 New Commitments

These are commitments that were adopted during the process of adjustments either because they are adjustments of input commitments, or because they were inferred from the system.

A.2.1 Adjusted Commitments

C 1 When there is a plausible threat of serious or irreversible harm to the environment, then uncertainty of the harm must not lead to postponing cost-effective measures to prevent it. [medium] [replaced IC 1 at Step B_3]

C 2 When an activity raises threats of harm to human health or the environment, then, *pro tanto*, precautionary measures should be taken even if some cause-and-effect relationships are not fully established scientifically. [medium] [replaced IC 2 at Step B_3]

C 3 Precautionary measures should not introduce threats that are equally or more severe than the threats they are aimed at, i.e., threats that have the same or a greater potential for harm. [medium] [replaced EC 15 at Step B_3]

C 4 The claim that the measles/mumps/rubella (MMR) vaccine might cause autism does not constitute a plausible threat. [medium] [replaced IC 12 at Step B_3]

C 5 The seriousness of a threat is assessed according to (i) the potential for harm of the threat, and (ii) whether or not this harm is incommensurable (e.g., because of being irreversible in some relevant sense) with other outcomes. [low] [replaced EC 3 at Step B4]

C 6 *Pro tanto*, every serious threat warrants precaution as long as it meets some minimal criteria of plausibility or reasonableness—i.e., that the likelihood of a possibility of severe harm is very low or cannot even be assigned is no *pro tanto* reason against taking precautions. [high] [replaced EC 4 at Step B4]

C 7 Threats to human health have lexical priority for taking precautionary measures insofar as they are threats of rights violations. [high] [replaced EC 23 at Step B5]

C 8 Threats to the environment have lexical priority for taking precautionary measures insofar as they are threats of rights violations. [high] [replaced EC 24 at Step B5]

C 9 Non-EC 1 [replaced EC 1 at Step B5]

C 10 In Case 9, *Job Offers*, the target system should not tell you to choose the job in New York. [high] [replaced IC 11 at Step B6]

C 11 The target system should not tell you not to wear protective clothing when making soap. [high] [replaced IC 9 at Step B6]

C 12 In Case 11, *Worst Case Being Shot*, the target system should not tell you to choose option A. [high] [replaced IC 10 at Step B6]

C 13 Not vaccinating your child is not a proportional precautionary measure against the alleged threat that the measles/mumps/rubella (MMR) vaccine might cause autism. [medium] [replaced C 4 as a replacement for IC 12 at Step B6]

A.2.2 *Newly Inferred Commitments*

These are commitments that have been inferred from the explication of "being a precautionary measure against an undesirable *x*" at Step B2 of the equilibration-process.

NC 1 Establishing a compensation scheme for future generations (IC 3) is a precautionary measure against the possibility that current mitigation and adaptation efforts will not be enough to prevent dangerous climate change.

NC 2 Requiring that any application of SRM against harmful impacts of climate change has to be accompanied by a strict mitigation and adaptation program (IC 6) is not a precautionary measure against other effects of increased GHG levels and negative effects from prolonged SRM-implementation. (Uncertainty about those negative effects is too low.)

NC 3 Requiring that SRM can only be applied if its international and intergenerational governance is clarified in a morally sound way (IC 7) is not a precautionary measure to prevent problems like those of responsibility, compensation (e.g., for damage and harm from changed weather patterns), distributive justice, and/or the termination problem. (Uncertainty about whether such problems would occur is too low.)

NC 4 In case 5, *Asbestos 1*, option (ii), starting systematic research, does count as a precautionary measure against the possibility that asbestos does cause lung disease and we discover it only very belated, and a lot of people suffer.

NC 5 In case 5, *Asbestos 1*, option (iii), research and strict regulation does count as a precautionary measure against the possibility that asbestos does cause lung diseases.

NC 6 In case 3, *R&D into SRM, two kinds of research*, choosing option (i), implementing a full-blown R&D program into SRM, does not count as a precautionary measure against dangerous climate change. (The (b)-aspect of the reasonableness-criterion is not fulfilled.)

NC 7 In case 3, *R&D into SRM, two kinds of research*, choosing option (ii), doing non-invasive research into SRM, especially the aspects that contribute to our general understanding of climate change, does count as a precautionary measure against dangerous climate change.

NC 8 In case 3, *R&D into SRM, two kinds of research*, choosing option (ii), doing non-invasive research into SRM, does count as a precautionary measure against the potential dangers of a full-blown R&D program into SRM.

NC 9 In case 3, *R&D into SRM, two kinds of research*, choosing option (iii), not implementing any research and/or development program into SRM, does count as a precautionary measure against the potential dangers of a full-blown R&D program into SRM.

NC 10 In case 3, *R&D into SRM, two kinds of research*, choosing option (iii), not implementing any research and/or development program into SRM, does count as a precautionary measure against the possibility that a non-invasive research program into SRM turns out to be a waste of money and effort that does not help us to prevent dangerous climate change.

NC 11 In case 6, *Asbestos 2*, choosing option (ii), research and regulation, is a precautionary measure against the possibility that people might get lung cancer from asbestos.

NC 12 In case 6, *Asbestos 2*, choosing option (iii), banning asbestos, is a precautionary measure against the possibility that people might get lung cancer from asbestos.

NC 13 In case 1, *Genetically Engineered Algae*, banning the technology is a precautionary measure against possible harmful effects from it.

NC 14 In case 9, *Job Offers*, choosing the job in Chicago is not a precautionary measure against having a tedious and badly paid job in New York. (By design of the case, it is certain that you end up with the bad job if you don't go to Chicago.)

NC 15 In case 9, *Job Offers*, choosing the job in New York is a precautionary measure against being killed in a plane accident.

NC 16 Not vaccinating your child with the MMR vaccine is not a precautionary measure against autism. [refers to IC 14]

NC 17 Vaccinating your child with the MMR vaccine is a precautionary measure to prevent your child from getting one of the diseases. [refers to IC 14]

NC 18 In case 5, *Asbestos 1*, option (iv), banning asbestos, does count as a precautionary measure against the possibility that asbestos does cause lung diseases.

NC 19 In case 6, Asbestos 2, choosing option (i), continuing business-as-usual, is neither a precautionary measure against the possibility that asbestos might cause lung cancer (would do nothing to prevent harm) nor against economic losses from restricting asbestos manufacturing (that there would be economic losses is too certain).

NC 20 Radiation therapy in cancer therapy is not a precautionary measure. (Uncertainty too low.) [refers to IC 17]

NC 21 In case 5, *Asbestos 1*, option (i), continuing business-as-usual, is neither a precautionary measure against the possibility that asbestos might cause lung diseases (would do nothing to prevent harm) nor against economic losses from restricting asbestos manufacturing (that there would be economic losses is too certain).

A.3 Candidate Systems

These are the various candidates for (parts of) the system that have been assessed and adjusted as part of the equilibration-process throughout Chaps. 6–8.

A.3.1 Rio PP and Wingspread PP

The Rio and Wingspread formulations of a precautionary principle (see IC 1 and IC 2) have been assessed as Principle 1 and Principle 2.

Principle 1 (P 1, The Rio PP) Where there are threats of serious or irreversible damage, lack of full scientific certainty [about those threats, T.R.] shall not be used as a reason for postponing cost-effective measures to prevent environmental degradation.

Principle 2 (P 2, The Wingspread PP) When an activity raises threats of harm to human health or the environment, precautionary measures should be taken even if some cause-and-effect relationships are not fully established scientifically.

A.3.2 The Principle 3-System (P 3)

Principle 3 (P 3) Where there are plausible threats of serious harm, precautionary measures that are proportional to the severity and plausibility of the threat should be taken.

P 3.1: Definition: Threat A threat is a possibility of harm that is uncertain.

P 3.2: Seriousness of Threats The seriousness of a threat is assessed according to (i) the potential for harm of the threat, and (ii) whether or not the possible harm is seen as reversible. [same content as IC 11]

P 3.3: *ExplicPrec* Explication of "Being a precautionary measure against an undesirable *x*": An action *a* is precautionary with respect to something undesirable *x* if *a* fulfills the following necessary and jointly sufficient criteria:

1. Intentionality: *a* is performed with the intention of preventing *x*.
2. Uncertainty: the agent does not believe it to be certain or highly probable that *x* will occur if *a* is not performed.
3. Reasonableness: the agent has externally good reasons (a) for believing that *x* might occur, (b) for believing that *a* will in fact at least contribute to the prevention of *x*, and (c) for not believing it to be certain or highly probable that *x* will occur if *a* is not performed.

A.3.3 The Rawlsian Core Precautionary Principle (RCPP)

Rawlsian Core Precautionary Principle (RCPP) *If* four conditions are fulfilled:

1. **No Probabilities:** There is no, or no reliable, probability information about the possible outcomes available,
2. **Care Little for Potential Gains:** decision-makers care relatively little for potential gains that might be made above the minimum that can be guaranteed by the maximin approach,

3. **Unacceptable Outcomes:** the courses alternative to the one selected by maximin have unacceptable outcomes, and
4. **Reasonable Outcomes:** the range of outcomes considered are in some appropriate sense "realistic" or reasonable,

then decision-makers should choose the course of action with the best worst case.

A.3.4 The Utilitarian Uncertainty Principle (UUP)

Utilitarian Uncertainty Principle (UUP) If no or no reliable probability information is available, treat all outcomes as equally probable, and choose the option that has the highest expected utility.

A.3.5 The Maximin-Precautionary Principle

Maximin-PP for Combinations of Uncertainty and Incommensurability (Maximin-PP) Select the course of action with the best worst case if you are either:

- In a situation of decision-theoretic risk or uncertainty (or some combination), and the outcomes of the available actions can be ranked on an ordinal scale, and all courses of action alternative to the one selected by maximin have outcomes that are incommensurably worse than the best worst case; or
- In a situation of (partial) decision-theoretic uncertainty, outcomes can be ranked on a cardinal scale, and all courses of action alternative to the one selected by maximin have negative outcomes that outweigh every potential gain that could be made above the level that can be guaranteed by maximin.

A.3.6 The Tripartite Precautionary Approach (TPA)

The **Tripartite Precautionary Approach (TPA)** consists of (cf. Steel 2015, 9–10):

- The **Meta Precautionary Principle (MPP)**: Uncertainty must not be a reason for inaction in the face of serious threats.
- The **Precautionary Tripod**: The elements that have to be specified in order to obtain an action-guiding precautionary principle version: If there is a threat that meets the *harm condition* under a given *knowledge condition* then a *recommended precaution* should be taken.

- **Proportionality**: Demands that the elements of the Precautionary Tripod are adjusted proportionally to each other, understood as *Consistency*: The recommended precaution must not be recommended against by the same PP version, and *Efficiency*: Among those precautionary measures that can be consistently recommended by a PP version, the least costly one should be chosen.
  The **strategy to obtain a PP decision rule by adjusting the precautionary tripod**: (1) select a desired safety target and define the harm condition as a failure to meet this target, (2) select the least stringent knowledge condition that results in a consistently applicable version of PP given the harm condition. To comply with the MPP, uncertainty must neither render the PP version inapplicable nor lead to continual delay in taking measures to prevent harm (cf. Steel 2015, 10).

A.3.6.1 Versions of the TPA

These are PP-versions obtained based on the TPA that were used when accounting for commitments. (Not all of them are proportional PP-versions.)

PP-Version 1 If there is a threat of unacceptable harm, which is a plausible outcome, then choose a decision option that does not have outcomes of unacceptable harms in its set of possible outcomes.

PP-Version 2 If there is a threat of catastrophic harm, which is a plausible outcome, then choose a decision option that does not have outcomes of catastrophic harms in its set of possible outcomes.

PP-Version 2* If there is a threat of catastrophic harm, which is a plausible outcome, then choose the decision option that has the best case.

PP-Version 3 If there is a threat of unacceptable harm, which is a plausible outcome, then choose the decision option with the best worst case.

PP-Version 4 If there is a threat of unacceptable harm, which is a plausible outcome, then choose the decision option with the highest cost-benefit balance.

PP-Version 5 If there is a threat of catastrophic harm, which is a plausible outcome, then choose the decision-option with the best worst case.

PP-Version 6 If there is a threat of death, which has a very small but positive probability, then take another action which does not have a very small but positive probability of death in its set of possible outcomes.

PP-Version 7 If there is an action that causes a threat of having a tedious and badly paid job, which is a certain outcome given the action, then choose an action that will have a better impact on overall quality of life.

PP-Version 7* If there is an action that causes a threat of having a tedious and badly paid job, which is a highly likely outcome given the action, then choose an action that will have a better impact on overall quality of life.

PP-Version 8 If there is a plausible possibility that a newly proposed activity would introduce a threat of serious harm to the environment and/or human health, then this activity should be prohibited by international law.

PP-Version 8* If there is a plausible possibility that an activity generates threats of serious harm to the environment and/or human health, then this activity should be prohibited.

A.3.7 *The Rights-Maximin-PP for Combinations of Uncertainty and Incommensurability*

Maximin-PP for Combinations of Uncertainty and Incommensurability (Maximin-PP) Select the course of action with the best worst case if you are either:

- In a situation of decision-theoretic risk or uncertainty (or some combination), and the outcomes of the available actions can be ranked on an ordinal scale, and all courses of action alternative to the one selected by maximin have outcomes that are incommensurably worse than the best worst case; or
- In a situation of (partial) decision-theoretic uncertainty, outcomes can be ranked on a cardinal scale, and all courses of action alternative to the one selected by maximin have negative outcomes that outweigh every potential gain that could be made above the level that can be guaranteed by maximin.

The Rights-Threshold Principle Threats of rights violations have lexical priority over other threats, and are incommensurable with chances of other kinds of gains.

A.3.8 *The Tripartite Precautionary Approach to Threats of Rights Violations*

The **Tripartite Precautionary Approach to Threats of Rights Violations (Rights-TPA)**:

- The **Rights Meta Precautionary Principle (Rights-MPP)**: Uncertainty must not be a reason for inaction when there are threats of rights implications.
- The **Precautionary Tripod**: The elements that have to be specified in order to obtain an action-guiding precautionary principle version: If there is a threat that meets the *harm condition* (i.e., a specific rights violation) under a given *knowledge condition* then a *recommended precaution* should be taken.
- **Proportionality**: Demands that the elements of the Precautionary Tripod are adjusted proportionally to each other, understood as *Consistency*: The recom-

mended precaution must not be recommended against by the same PP version, and *Efficiency*: Among those precautionary measures that can be consistently recommended by a PP version, the least costly one should be chosen.
The **starting point for a rights-based PP version**: If there is (1) a threat of a wrongful rights violation, then (2) select the least stringent knowledge condition that results in a consistently applicable version of PP given the harm condition. To comply with the Rights-MPP, uncertainty must neither render the PP version inapplicable nor lead to continual delay in taking measures to prevent rights violations.

A.4 Background

The background cannot be described exhaustively. Some relevant parts are identified in Chap. 5, Sect. 5.4. In the following, I provide the full case descriptions of all the used toy examples.

A.4.1 Case Descriptions

Case 1: Genetically Engineered Algae "[A] breakthrough has been achieved in genetic engineering. Ways have been found to control and modify the metabolism of a species of microalgae with unprecedented ease. "Synthesizing a chemical with this technology is more like programming a computer than modifying an organism," said one of the researchers. A group of critics demand that the new technology be prohibited by international law. They point to its potential dangers, such as the spread of algae that produce highly toxic substances" (Hansson 2016, 95). While the basic mechanisms of the technology are well understood, there has not been much research yet into possible side-effects etc. It is indeed a plausible threat that this technology might lead to unintended negative side-effects, such as the spread of algae that produce highly toxic substances. However, it is equally plausible that the new technology will help to solve important problems of humanity. Cheap production of important medicine, renewable energy production, or something similar, has been shown to be at least as plausible as the harmful effects. There are currently no other means available that are equally promising (Fig. A.1).

Case 2: R&D into SRM A strict mitigation and adaptation policy is implemented, but dangerous climate change is still possible because of feedback effects and tipping points. There are no signs that a catastrophe is imminent in the next 5 years. The basic mechanisms of solar radiation management are known, but there are still huge uncertainties, e.g. about its effects on a local level and possible (so far unforeseen, possibly catastrophic) side-effects. Should we do research and development on SRM with the goal of developing it ready to use?

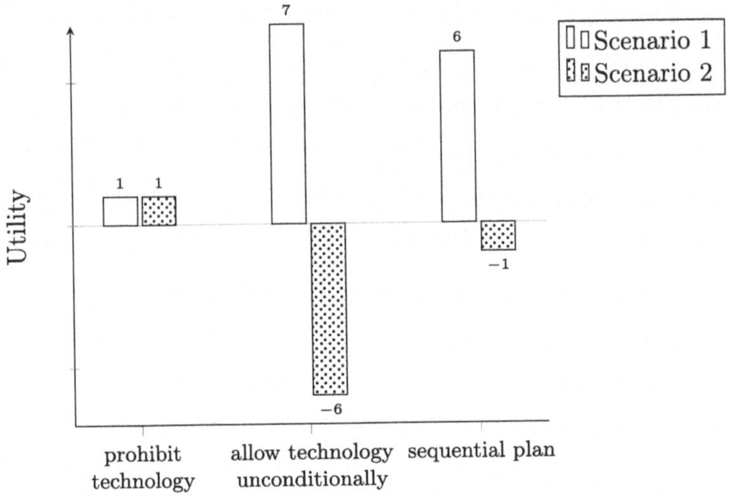

Fig. A.1 Possible Outcomes in Case 1, *Genetically Engineered Algae*

We know that:[2] R&D has no, neither positive nor negative, influences on our mitigation and adaptation efforts, and that R&D itself does not pose any additional threats to the climate system.

We are given two choices: (i) implementing a research and development (R&D) program for SRM with the objective of developing SRM ready to use, or (ii) not implementing an R&D program for SRM.

Case 3: R&D into SRM, Two Kinds of Research A strict mitigation and adaptation policy is implemented, but dangerous climate change is still possible because of feedback effects and tipping points. There are no signs that a catastrophe is imminent in the next 5 years. The basic mechanisms of solar radiation management are known, but there are still huge uncertainties, e.g. about its effects on a local level and possible (so far unforeseen, possibly catastrophic) side-effects. Should we do research and development (R&D) into SRM in order to develop it ready to use?

We know that full-blown R&D into SRM introduces threats of its own, since it makes field-experiments necessary that already involve implementation of SRM technology. On the other hand, without these field tests, substantial uncertainties about SRM cannot be reduced. The outcomes of our alternative options will depend on whether or not climate sensitivity is high, and whether or not field tests will have unforeseen negative side effects (that, in the worst outcome, could amount to a climate change catastrophe).

[2] Obviously, these are simplifications which are not realistic.

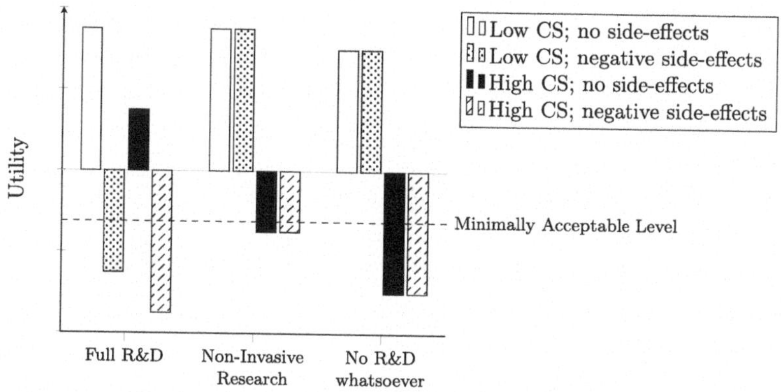

Fig. A.2 Possible Outcomes in Case 3, *R&D into SRM, Two Kinds of Research*

We are given three choices:

(i) implementing a full-blown research and development (R&D) program into SRM with the objective of developing SRM ready to use,

(ii) doing non-invasive research into SRM, especially the aspects that contribute to our general understanding of climate science, or

(iii) not implementing an R&D program for SRM.

The outcomes of this choices depend on the climate sensitivity, and whether or not field tests will have unforeseen side-effects, i.e., like "SRM does not work" (Fig. A.2).

Case 5: Asbestos 1 Large-scale mining and manufacturing of asbestos has started about 15 years ago. Asbestos is seen as a desirable material because of its properties like sound absorption, tensile strength, and its resistance to fire and heat. Production costs are low, so it is also affordable. However, there are observations and reports that associate lung diseases with inhaling asbestos, although no systematic scientific research has been done on it so far; thus, a clear connection cannot be proved, and the diseases might have other causes.

We have to choose between the following four options:

(i) BAU: Continuing business-as-usual,

(ii) Research: Starting systematic scientific research on the harmfulness of asbestos dust, including long-term studies and mortality statistics of asbestos workers,

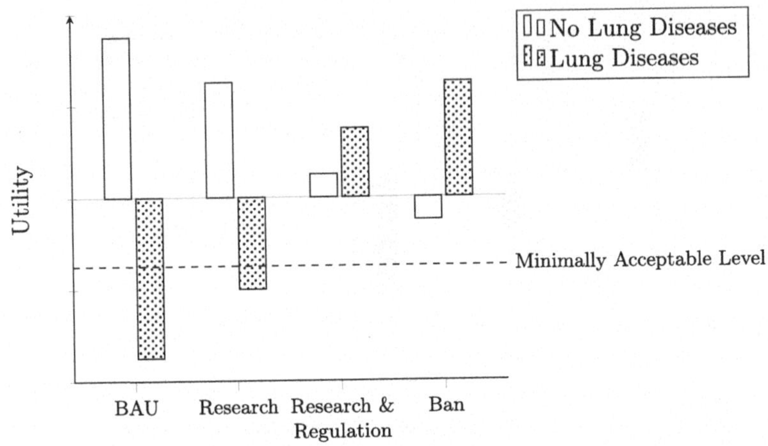

Fig. A.3 Possible Outcomes in Case 5, *Asbestos 1*

(iii) Research&Regulation: Starting systematic scientific research while already strictly regulating asbestos production, including, e.g., limiting exposure of workers to asbestos dust, and making compensation arrangements, based on agreed liabilities, or

(iv) Ban: Banning asbestos (Fig. A.3).

Case 6: Asbestos 2 Large-scale mining and manufacturing of asbestos has started about 45 years ago. Asbestos is seen as a desirable material because of its properties like sound absorption, tensile strength, and its resistance to fire and heat. Production costs are low, so it is also affordable. It is widely used in a range of applications, and its use is continuing to grow. However, it is now accepted that the inhalation of asbestos dust can cause a lung disease called "asbestosis".[3] Recently there have been cases of asbestosis that have been complicated by lung cancer, but a clear connection is difficult to prove, one reason being that smoking has become increasingly popular and is also seen as a potential cause for lung cancer.[4] Additionally, some concerns have been raised that the inhalation of asbestos dust might cause other long-latent-period harm to people. There are other, presumably safer substances available, but they are much more expensive in production costs.[5]

[3] E.g., a health study of asbestos workers has shown that 66% of those employed for 20 years or more suffered from asbestosis, versus none of those employed for less than 4 years (Harremoës et al. 2001, 54).

[4] I omit here that in Germany, before smoking became popular and while lung cancer rates were still relatively low, a connection between asbestos and lung cancer was already accepted in 1938 (Harremoës et al. 2001, 54).

[5] For reasons of simplicity, I do not consider what kinds of measures were already taken, and how effective (or not) they have been.

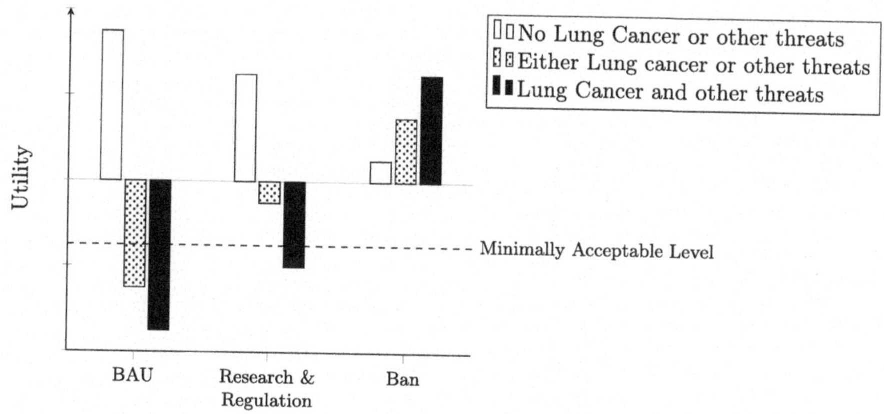

Fig. A.4 Possible Outcomes in Case 6, *Asbestos 2*

We have to choose between he following three options:

(i) BAU: Continuing business-as-usual,

(ii) Research&Regulation: Starting systematic scientific research while already strictly regulating asbestos production, including, e.g., limiting exposure of workers to asbestos dust, and making compensation arrangements, based on agreed liabilities,

(iii) Ban: Banning asbestos (Fig. A.4).

Case 7: Disproportionate Outcomes 1 We have to decide between two alternative courses of action, "Safe 1" and "Risky 1" , see Fig. A.5.

Case 8: Disproportionate Outcomes 2 We have to decide between two alternative courses of action, "Safe 2" and "Risky 2" , see Fig. A.6.

Case 9, Job Offers Suppose you live in New York City and are offered two jobs at the same time. One is a tedious and badly paid job in New York City itself, while the other is a very interesting and well-paid job in Chicago. But the catch is that, if you wanted the Chicago job, you would have to take the plane from New York City to Chicago (e.g., because this job would have to be taken up the very next day). Therefore there would be a very small but positive probability that you might get killed in a plane accident (example from Harsanyi 1975, 595).

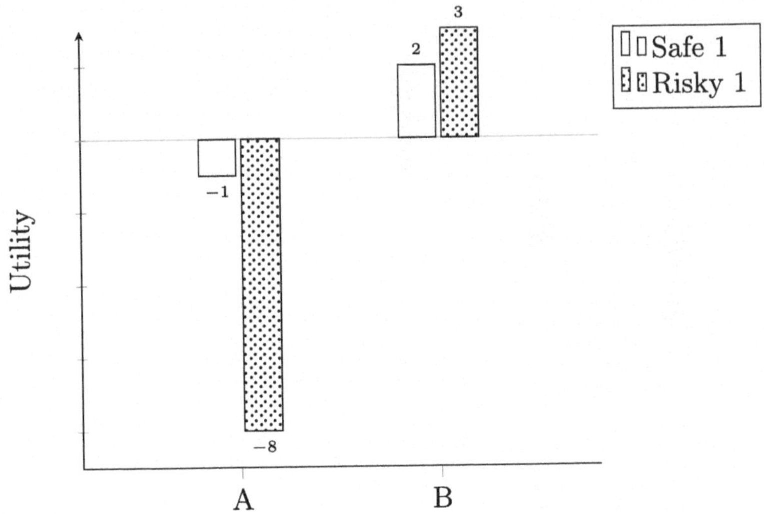

Fig. A.5 Possible Outcomes in Case 7, *Disproportionate Outcomes 1*

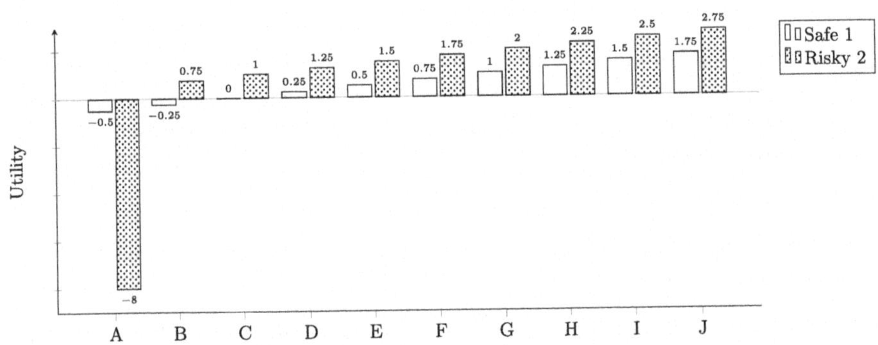

Fig. A.6 Possible Outcomes in Case 8, *Disproportionate Outcomes 2*

Case 10, Job Offers 2 A reporter, living in Los Angeles, has been told that he can take one of two assignments. First, he can go to a nation, say Iraq, that is facing a large amount of terrorism. Second, he can go to Paris to cover anti-American sentiment in France. The Iraq assignment has, in his view, two polar outcomes: (a) he might have the most interesting and rewarding experience of his professional life or (b) he might be killed. The Paris assignment has two polar outcomes of its own: (a) he might have an interesting experience, one that is also a great deal of fun and (b) he might be lonely and homesick (example from Sunstein 2007, 12).

Case 11, Worst Case being Shot Suppose that in a given situation you have two actions, A and B, available to you. If you choose A, then there are two possible outcomes: either (A1) you will receive $100, or (A2) you will be shot. If you choose B, there are also two possible outcomes: Either (B1) you will receive $50, or (B2) you will receive a slap on the wrist (example from Gardiner 2006, 45–46).

Case 12: Chemical Waste "A company applies for an emission permit to discharge its chemical waste into an adjacent, previously unpolluted lake. The waste in question has no known ecotoxic effects. A local environmental group opposes the application, claiming that the substance may have unknown deleterious effects on organisms in the lake.

[...] We know from experience that chemicals can harm life in a lake, but we have no correspondingly credible reasons to believe that a chemical can improve the ecological situation in a lake. (To the extent that this "can" happen, it does so in a much weaker sense of "can" than that of the original argument [...])." (Hansson 2016, 96)

A.5 Schematic Overview of the Process of Adjustments

Figure A.7 gives a schematic overview of the process of adjustments and its three phases (which correspond to Chaps. 6–8). B-Steps, in which the commitments are adjusted, are on the left side, and A-Steps, in which the system is adjusted, are on the right side. The process starts from the set of initial commitments (IC), C_0. Candidates or commitments that were rejected at a given step are crossed out. If a commitment was replaced with a new commitment, this is indicated by a small arrow. When emerging commitments (EC) were made explicit at a specific step, or commitments that followed from the system were adopted as newly inferred commitments (NC), this is indicated by a plus sign.

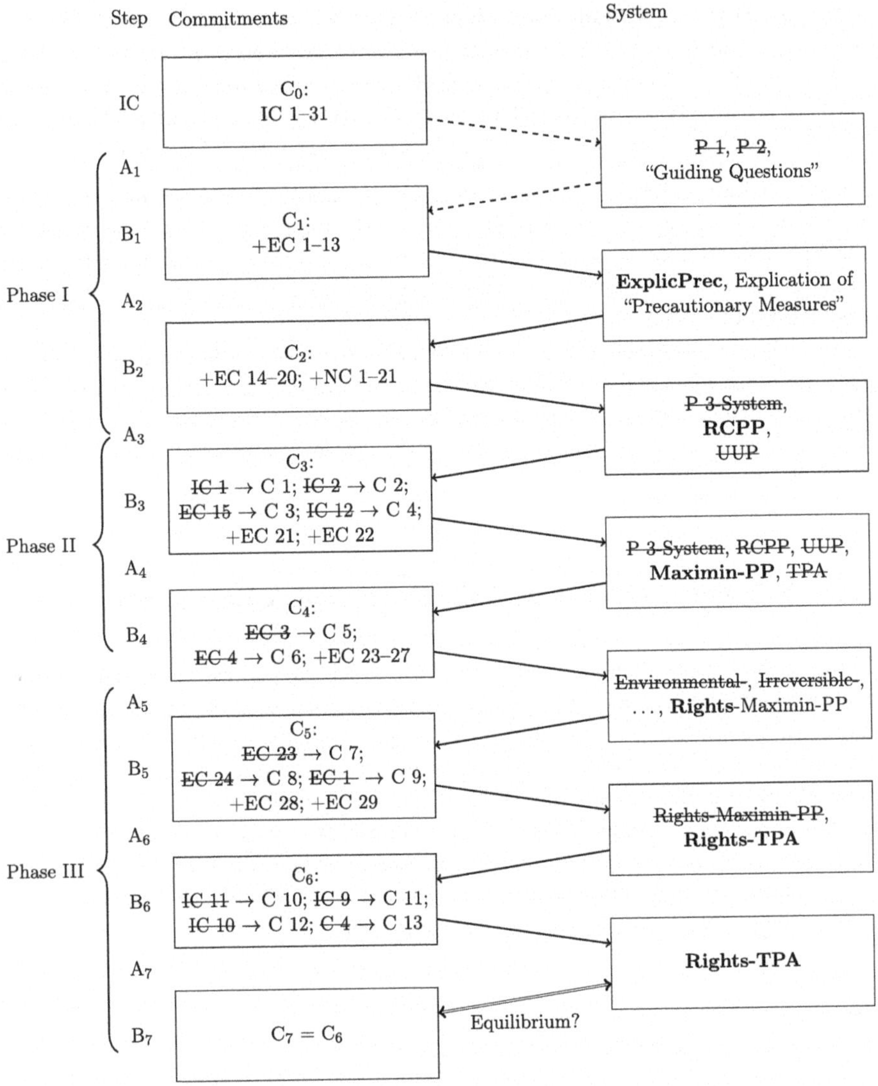

Fig. A.7 Schematic overview of the RE-Application and its three phases

References

Gardiner SM (2006) A Core precautionary principle. J Polit Philos 14(1):33–60

Hansson SO (2016) Evaluating the uncertainties. In: Hansson SO, Hirsch Hadorn G (eds) The argumentative turn in policy analysis, Springer, Berlin, pp 79–104

Harremoës P, Gee D, MacGarvin M, Stirling A, Keys J, Wynne B, Vaz SG (eds) (2001) Late lessons from early warnings: the precautionary principle 1896–2000. Office for Official Publications of the European Communities, Luxembourg

Harsanyi JC (1975) Can the maximin principle serve as a basis for morality? A critique of John Rawls's Theory. Am Polit Sci Rev 69(2):594–606. https://doi.org/10.2307/1959090

Sandin P (2004) The precautionary principle and the concept of precaution. Environmental Values 13(4):461–475

Steel D (2015) Philosophy and the Precautionary Principle. Cambridge University Press, Cambridge, UK

Sunstein CR (2007) The catastrophic harm precautionary principle. Issues in Legal Scholarship 6(3):1–29

Index

T. Rechnitzer, *Applying Reflective Equilibrium*, Logic, Argumentation & Reasoning 27, https://doi.org/10.1007/978-3-031-04333-8

The manufacturer's authorised representative in the EU is Springer Nature Customer Service Centre GmbH, Europaplatz 3, 69115 Heidelberg, Germany. If you have any concerns regarding our products, please contact ProductSafety@springernature.com

Printed and bound by CPI Group (UK) Ltd, Croydon, CR0 4YY
22/09/2026
02220982-0012